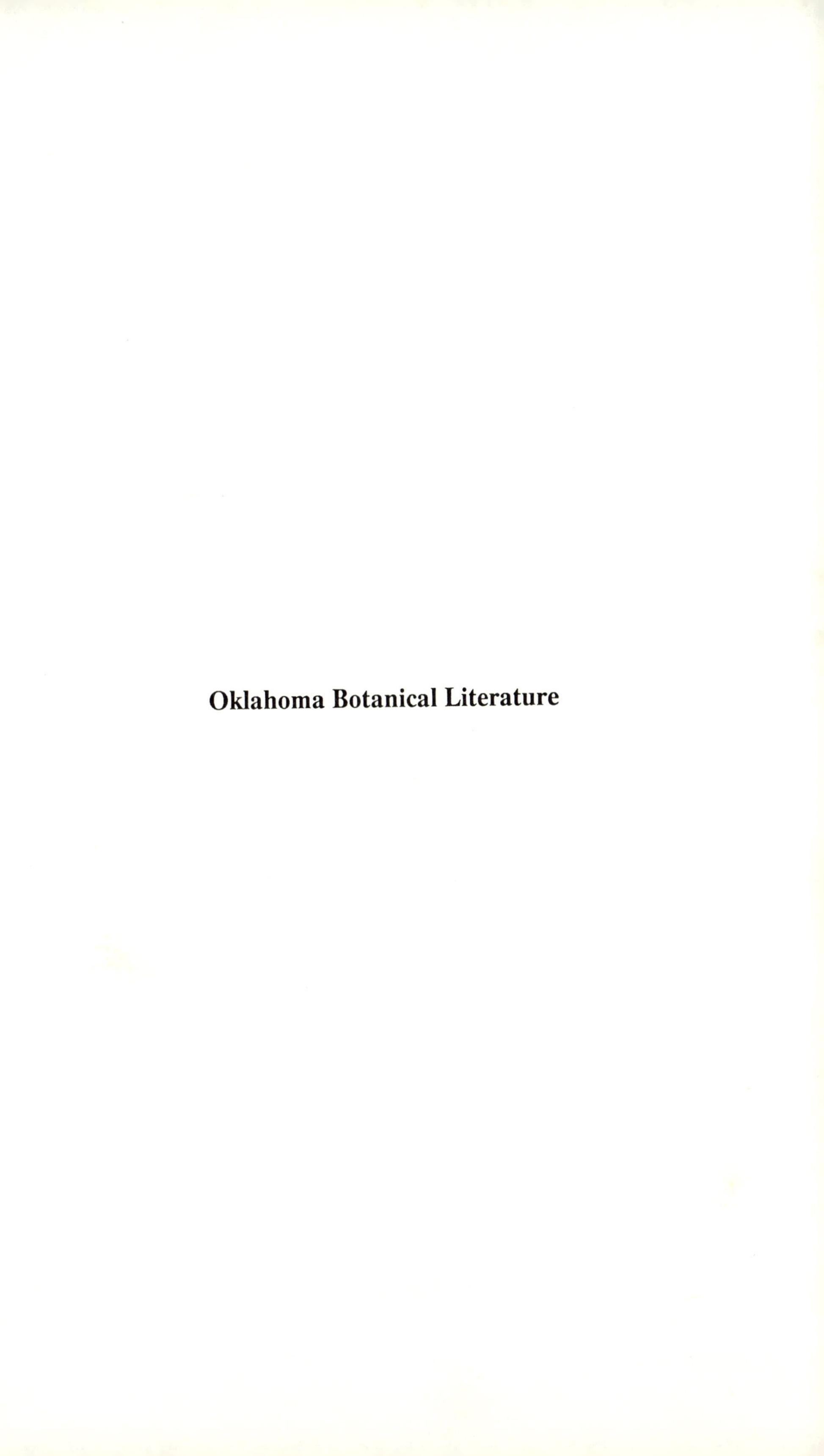

Oklahoma Botanical Literature

Oklahoma Botanical Literature

Forrest L. Johnson and T. H. Milby

University of Oklahoma Press: Norman and London

Oklahoma Botanical Literature is published by the University of Oklahoma Press for the Oklahoma Biological Survey, University of Oklahoma, Norman.

Cover and title page drawings of *Buchloe dactyloides* by T. H. Milby.

Library of Congress Cataloging-in-Publication Data

Johnson, Forrest L.
Oklahoma botanical literature / Forrest L. Johnson and T. H. Milby – 1st ed.
p. cm.
"Published ... for the Oklahoma Biological Survey, University of Oklahoma, Norman" – P.
Includes indexes.
ISBN 0-8061-2198-X (alk. paper)
1. Botany – Oklahoma – Bibliography. I. Milby, T. H. (Torbert Hickman), 1926- . II. Title.
Z5358.U5054 1989
[QK181]
016.581'09766 – dc19 88-31247

The paper in this book meets the guidelines for permanence and durability of the Committee on Production Guidelines for Book Longevity of the Council on Library Resources, Inc.

CONTENTS

Introduction	3
Literature Citations	5
Author Index	63
County Index	69
Section-of-State Index	70
Duck and Fletcher Type Index	71
Other Vegetation Type Index	71
Subject-in-Context Index	72

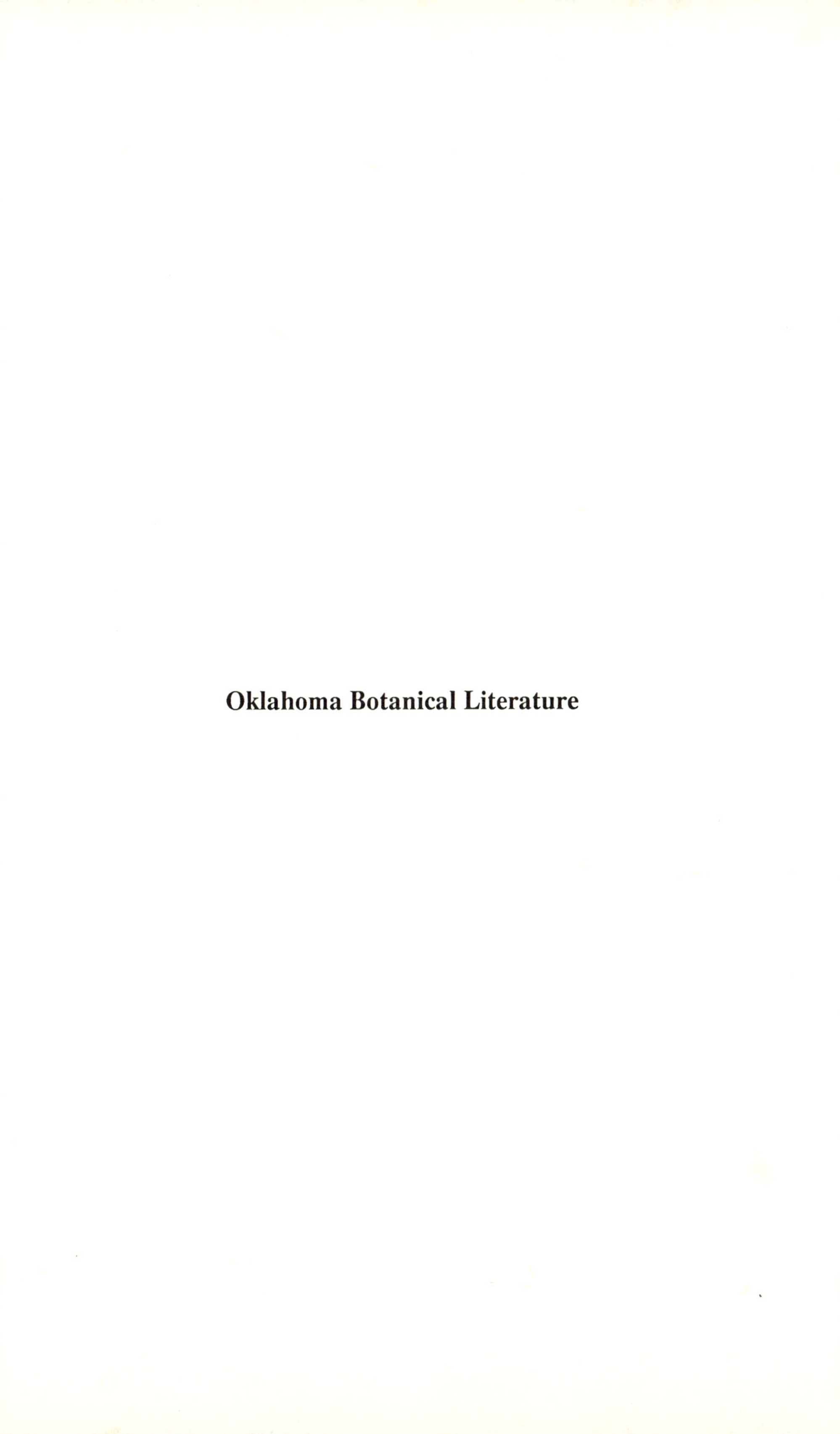

Oklahoma Botanical Literature

INTRODUCTION

The journal of Thomas Nuttall describing his travels in the year 1819 in the western part of Arkansas Territory gives us the earliest report of the plants of the region that was to become the state of Oklahoma. The publication of Nuttall's flora (1837) and that of John Torrey (1828) are the first documented collections of the flora of this heretofore unexplored territory. Military explorations by J. W. Abert (1846), Nathan Boone (1843), Edwin James (1823), R. B. Marcy (1850, 1853), Thomas Wheelock (1834), A. W. Whipple (1855), and others, each of which provided descriptions of plants and other natural elements of the region, led the way to what has become in the century since settlement a large and diverse literature on the plants of Oklahoma. Even our first recorded tourist, Washington Irving (1835), provided some good observations concerning the presettlement vegetation.

The purpose of this bibliography of botanical works pertaining to Oklahoma is to identify and bring together this widely dispersed literature as an aid to further research and study. It includes works that have been published through 1987. More than 850 published works have been included in the bibliography. This compilation brings the work of Ralph Kelting and W. T. Penfound (1953), T. H. Milby and W. T. Penfound (1965), and T. H. Milby (1977) up to date. An extensive index was developed that makes the present bibliography easier to use than the earlier versions.

Numerous additions to the botanical literature of Oklahoma will be made in the future. To accommodate updating and revising this bibliography, we have stored and processed it on a microcomputer at the Oklahoma Biological Survey using dBASE III™, a general-purpose storage-and-retrieval system. Thus, the bibliography can be easily sorted and searched on the basis of any information contained in the included citations.

Typesetting was done on the same microcomputer that contains the database. We employed Wordstar 2000+™ software and a Hewlett-Packard Laserjet Series II™ printer to produce camera-ready copy.

KINDS OF MATERIAL INCLUDED IN THE LIST

Most publications listed are primary sources such as journal articles, masters' theses, and doctoral dissertations that report original research on various aspects of plant life in Oklahoma. We also included a number of secondary sources describing botanical elements of the state. Several regional and national floras and other descriptive works that give good coverage of Oklahoma are included. Floras that pertain to the plants of the state as a whole or to those within a specifically defined subdivision or region of the state are included along with taxonomic descriptions of plants that were originally discovered and described in the state. Some agricultural publications that deal with soil conservation or with native plants are included, but those dealing with crops are not. Publications on algae and fungi of Oklahoma soil and water are within the purview of the work, but those having to do with fossil plants and palynology are not. Zoological, geographical, and geological publications that contain vegetation descriptions are included. Certain special reports to and by government agencies are included that deal with Oklahoma vegetation and are readily obtainable. Publications that deal exclusively with anatomy, morphology, or physiology of plants are not included. Numerous short notes concerning range extensions and other taxonomic matters were omitted because the details usually appeared in

longer publications at a later date. Many environmental impact statements describe vegetation, but we have found that they are generally unavailable except for a short time after publication and have omitted most of them from this list.

HOW THE MATERIAL IS ARRANGED

Entries in the bibliography are arranged in alphabetical order by the last name of the principal author or by the name of the organization or agency under whose auspices the studies were undertaken and the results published, e.g., U.S. War Department (1855), U.S. Geological Survey (1926). Several of the journals of early military expeditions into the area are listed under both the author and the sponsoring government agency. Each entry consists of: citation number, author(s), year, title, source, and additional keywords {in brackets}. All authors are included in an alphabetical author index. Abbreviations of journal titles follow the BIOSIS standard of *Biological Abstracts*.

HOW THE MATERIAL WAS COLLECTED

Major North American as well as state and regional botanical journals were examined to discover articles pertaining to Oklahoma that should be included. The catalogs of the libraries at Oklahoma State University and the University of Oklahoma were searched for theses and dissertations relating to Oklahoma botany as well as those from all other universities in the state at which graduate degrees in biology may have been granted. Botanists from all the universities in Oklahoma were solicited to contribute their own personal bibliographies of botanical publications for inclusion.

INDEXES

Material in the bibliography can be found through indexes that are arranged by author, county, vegetation type, section-of-state, and keyword (subject). The section-of-state index is based on the seven sections defined by Kelting and Penfound (1953).

ACKNOWLEDGMENT

We wish to thank the following people who have contributed material for inclusion in the bibliography: P. Buck, S. L. Collins, J. J. Crockett, E. E. Dale, Jr., J. R. Estes, G. J. Goodman, C. A. Lawson, L. K. Magrath, J. K. McPherson, L. A. Pfiester, J. J. Skvarla, C. E. Taylor, R. J. Taylor, R. J. Tyrl, G. E. Uno, and L. L. Wallace.

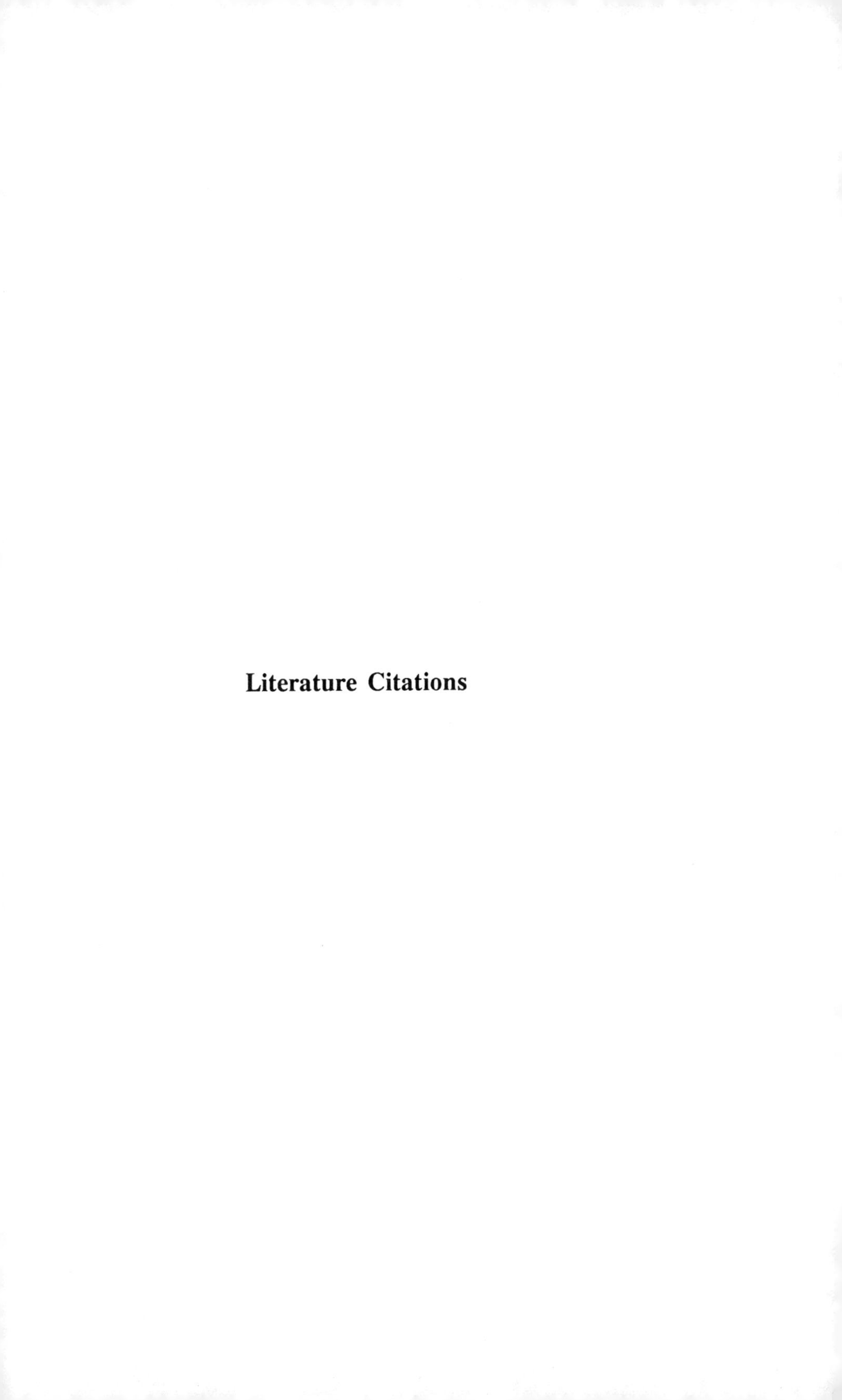

Literature Citations

LITERATURE CITATIONS

1. **Abbott, J. E.** 1974. Introgressive hybridization between shortleaf and loblolly pine in southeastern Oklahoma. M.S. thesis, Okla. State Univ., Stillwater. {SE forest genetics}

2. **Abdul-Wahab, A. S.** 1967. Plant inhibition by Johnson grass and its possible significance in plant succession. Ph.D. diss., Univ. Okla., Norman. {allelopathy grassland}

3. **Abdul-Wahab, A. S., and E. L. Rice.** 1967. Plant inhibition by Johnson grass and its possible significance in old field succession. Bull. Torrey Bot. Club 94: 486-497. {allelopathy grassland}

4. **Abert, J. W.** 1846. Journal of Lt. J. W. Abert from Bent's Fort to St. Louis in 1845. U.S. 29th Cong., 1st Sess., Sen. Doc. 8, No. 438: 1-75. (see: U.S. Senate) {presettlement}

5. **Adams, D. B.** 1971. Some factors influencing the frequency of bark lichens in north central Oklahoma. M.S. thesis, Univ. Okla., Norman. {NC forest climate}

6. **Adams, D. B., and P. G. Risser.** 1971. Some factors influencing the frequency of bark lichens in north central Oklahoma. Amer. J. Bot. 58: 752-757. {NC upland forest climate}

7. **Adams, D. B., and P. G. Risser.** 1971. The effect of host specificity on the interspecific associations of bark lichens. Bryologist 74: 451-457. {NC upland forest}

8. **Adams, D. E.** 1982. Nutrient and biomass allocation strategies for five grass species in an Oklahoma tallgrass prairie. Ph.D. diss., Univ. Okla., Norman. {grassland nutrient}

9. **Adams, D. E., R. C. Anderson, and S. L. Collins.** 1982. Differential response of woody and herbaceous species to summer and winter burning in an Oklahoma prairie. Southw. Natur. 27: 55-61. {fire grassland SC Marshall}

10. **Adams, D. E., W. E. Perkins, and J. R. Estes.** 1981. Pollination systems in *Paspalum dilatatum* Poir. (Poaceae): An example of insect pollination in a temperate grass. Amer. J. Bot. 68: 389-394. {grass SC Marshall}

11. **Adams, D. E., and L. L. Wallace.** 1985. Nutrient and biomass allocation in five grass species in an Oklahoma tallgrass prairie. Amer. Midl. Natur. 113: 170-181. {N P K SC grassland nutrient}

12. **Adams, D. E., and R. C. Anderson.** 1978. The response of a central Oklahoma grassland to burning. Southw. Natur. 23: 623-632. {NC grassland fire}

13. **Ahshapanek, D. C.** 1959. Phenology of a native tall-grass prairie in central Oklahoma. Ecology 43: 135-139. {SC grassland McClain}

14. **Ahshapanek, D. C.** 1959. Phenology of a native tall-grass prairie in central Oklahoma. M.S. thesis, Univ. Okla., Norman. {SC grassland McClain}

15. **Ahshapanek, D. C.** 1962. Ecological studies of plant inhibition by *Solanum rostratum*. Ph.D. diss., Univ. Okla., Norman. {allelopathy}

16. **Al Naib, F. A. G.** 1968. Allelopathic effects of *Platanus occidentalis* L. M.S. thesis, Univ. Okla., Norman. {SC Cleveland allelopathy sycamore}

17. **Al Naib, F. A. G.** 1970. Allelopathic effects of *Platanus occidentalis* L. in relation to the patterning of vegetation. Ph.D. diss., Univ. Okla., Norman. {SC Pottawatomie allelopathy sycamore}

18. **Al Naib, F. A., and E. L. Rice.** 1971. Allelopathic effects of *Platanus occidentalis*. Bull. Torrey Bot. Club 98: 75-82. {SC allelopathy}

19. **Al Saadawi, I. S.** 1981. Allelopathic effects of *Polygonum aviculare* L. Ph.D. diss., Univ. Okla., Norman. {SC Cleveland}

20. **Albertson, F. W., and J. E. Weaver.** 1945. Injury and death or recovery of trees in a prairie climate. Ecol. Monogr. 15: 393-433. {}

21. **Allameh, A.** 1979. The effect of cement-kiln dust on vegetation. M.S. thesis, Univ. Okla., Norman. {SC forest pollution}

22. **Allan, P. F., and B. Osborn.** 1954. Tall grass defeats prairie dogs. Soil Cons. 20:103-105. {SW grassland prairie}

23. **Alleman, D. R.** 1982. Comparative water relations of four species of upland deciduous trees along a moisture gradient. M.S. thesis, Okla. State Univ., Stillwater. {NC forest}

24. **Allgood, F. P., and F. Gray.** 1974. Ecological interpretation for small mounds in landscapes of eastern Oklahoma. J. Environ. Qual. 3: 37-41. {SE mima pimple mounds}

25. **Allred, B. W.** 1949. Distribution and control of several woody plants in Texas and Oklahoma. J. Range Manage. 2: 17-29. {grassland prairie savannah management}

26. **Amos, B. B.** 1981. Reproductive studies in the genus *Callirhoe* (Malvaceae). Ph.D. diss., Univ. Okla., Norman. {flora}

27. **Anderson, K. L., and C. L. Fly.** 1955. Vegetation-soil relationships in Flint Hills bluestem pastures. J. Range Manage. 8: 163-169. {NC prairie grassland}

28. **Anderson, R. C., and D. E. Adams.** 1981. Flowering patterns and production in a central Oklahoma grassland. Ohio Biol. Surv. Notes No. 15. {prairie phenology NC}

29. **Anonymous.** 1963. Pasture and range plants. Phillips Petrol. Co., Bartlesville, OK. {prairie grassland all}

30. **Arter, B. S.** 1984. Growth and distribution of *Platanus occidentalis* in Oklahoma as influenced by climate and water relations. M.S. thesis, Okla. State Univ., Stillwater. {forest bottomland}

31. **Atkins, A. P.** 1950. Conservation ranching in the Oklahoma panhandle. J. Range Manage. 3: 167-171. {PH grassland management}

32. **Atwood, J. T.** 1985. The range of *Cypripedium kentuckiense*. Amer. Orchid Soc. Bull. 54(10): 1197-1199. {flora orchid}

33. **Baalman, R. J.** 1965. Vegetation of the Salt Plains Wildlife Refuge, Jet, Oklahoma. Ph.D. diss., Univ. Okla., Norman {NW Alfalfa salt}

34. **Baldock, L. O.** 1938. Flora of Kiowa County, Oklahoma. M.S. thesis, Okla. A & M Coll., Stillwater. {SW flora}

35. **Barber, S. C.** 1975. A floristic study of the vascular plants of the Gypsum Hills and Redbed Plains area of southwestern Oklahoma. M.S. thesis, Okla. State Univ., Stillwater. {SW flora}

36. **Barber, S. C.** 1979. Floristic components of the Gypsum Hills and Redbed Plains of southwestern Oklahoma. Southw. Natur. 24: 431-437. {SW Harmon Jackson Greer flora}

37. **Barber, S. C.** 1980. Taxonomic studies in the *Verbena stricta* complex. Ph.D. diss., Univ. Okla., Norman. {flora}

38. **Barber, S. C., and J. R. Estes.** 1978. Comparative pollination ecology of *Pyrrhopappus geiseri* and *Pyrrhopappus carolinianus*. Amer. J. Bot. 65: 562-566. {SC Marshall Cleveland Texas}

39. **Barclay, H. G.** 1938. A preliminary report on the ecology of a *Drosera* meadow. Proc. Okla. Acad. Sci. 18: 22-25. {SE sundew}

40. **Barclay, H. G.** 1947. The woody vegetation of Bear's Glen, a Washington Irving stopover. Proc. Okla. Acad. Sci. 28: 40-57. {NC upland forest tree phytosociology}

41. **Barclay, H. G.** 1970. The Nature Conservancy's first Oklahoma preserve. Proc. Okla. Acad. Sci. 50: 125. {NE}

42. **Barker, B. M., and W. C. Jameson.** 1975. Platt National Park: Environment and ecology. Univ. Okla. Press, Norman. {SC Murray}

43. **Barkley, E. A.** 1933. A preliminary survey of the vascular plants of Pottawatomie County, Oklahoma. Proc. Okla. Acad. Sci. 14: 44-46. {flora SC}

44. **Barkley, E. A. D.** 1933. A preliminary list of the vascular plants of Pottawatomie County, Oklahoma. M.S. thesis, Univ. Okla., Norman. {flora SC}

45. **Barkley, E. A., and C. C. Smith.** 1934. A preliminary study of buffalo wallows in the vicinity of Norman, Oklahoma. Proc. Okla. Acad. Sci. 14: 47-52. {SC Cleveland prairie grassland}

46. **Baumgartner, F. M., and A. M. Baumgartner.** 1987. *Lilium michiganense* Farwell (Liliaceae) and *Gentiana alba* Muhl. (Gentianaceae) new to Oklahoma. Sida 12: 249. {flora}

47. **Beare, M. H.** 1983. Water quality and phytoplankton communities in Lake Thunderbird, Cleveland County, Oklahoma. M.S. thesis, Univ. Okla., Norman. {aquatic algae SC}

48. **Becker, D. A., and J. J. Crockett.** 1973. Evaluation of sampling techniques on tall-grass prairie. J. Range Manage. 26: 61-65. {NC Payne grassland tallgrass method}

49. **Bessinger, G. H.** 1973. The influence of varying light intensities on the vegetative growth of ten southern forest deer browse species. M.S. thesis, Okla. State Univ., Stillwater. {forest}

50. **Bigham, H. R.** 1970. The comparative utility of plant life-form and other vegetative characters in evaluating the habitat of the cottontail. Ph.D. diss., Okla. State Univ., Stillwater. {NC}

51. **Birdsey, R. A., and D. F. Bertelson.** 1987. Forest statistics for southeast Oklahoma counties – 1986. USDA Forest Ser. Resource Bull. SO-119. {SE McCurtain LeFlore Pushmataha Choctaw Latimer Haskell Atoka Bryan}

52. **Birdsey, R. A., and D. F. Bertelson.** 1987. Forest statistics for northeast Oklahoma – 1986. USDA Forest Ser. Resource Bull. SO-120. {NE Sequoyah Adair Delaware Ottawa Mayes Cherokee Muskogee McIntosh}

53. **Blair, A. P.** 1978. New plant distribution records in eastern Oklahoma. Proc. Okla. Acad. Sci. 58: 106-107. {NE SE flora}

54. **Blair, W. F.** 1938. Ecological relationships of the Bird Creek region, northeastern Oklahoma. Amer. Midl. Natur. 20: 473-526. {NC grass grassland NE}

55. **Blair, W. F., and T. H. Hubbell.** 1938. The biotic districts of Oklahoma. Amer. Midl. Natur. 20: 425-455. {classification all}

56. **Blinn, W. C.** 1951. The short grass plains and post oak-blackjack woodland of Oklahoma in historical perspective. M.S. thesis, Okla. A & M Coll., Stillwater. {grassland prairie forest}

57. **Blum, U.** 1968. Inhibition of symbiotic nitrogen-fixation by gallic and gallotannic acids and possible roles in old-field succession. Ph.D. diss., Univ. Okla., Norman. {allelopathy}

58. **Blum, U., and E. L. Rice.** 1969. Inhibition of symbiotic nitrogen fixation by gallic and tannic acid, and possible roles in old field succession. Bull. Torrey Bot. Club 96: 531-544. {allelopathy}

59. **Bogue, E. E.** 1900. Annotated catalog of the ferns and flowering plants of Oklahoma. Okla. Exp. Sta. Bull. No. 45: 3-48. {flora all}

60. **Bogue, E. E.** 1900. Native Oklahoma plants. Okla. Agric. Exp. Sta. Bull. No. 45. {flora all}

61. **Bokhari, V. G., and J. S. Singh.** 1975. Standing state and cycling of nitrogen in soil-vegetation components of prairie ecosystems. Ann. Bot. 39: 273-285. {NC Osage grassland N nutrient}

62. **Bol, E. L.** 1984. Biosystematics of *Artemisia carruthii* and *Artemisia mexicana*. M.S. thesis, Univ. Okla., Norman. {flora}

63. **Bollinger, C. J.** 1925. The eastern boundary of the Great Plains in north central Oklahoma. Proc. Okla. Acad. Sci. 5:123-124. {NC classification}

64. **Boone, N.** 1843. Captain Nathan Boone's journal. Chron. Okla. 7: 58-105. {presettlement}

65. **Booth, W. E.** 1932. The relation of plant communities to soil erosion in Cleveland County, Oklahoma. M.S. thesis, Univ. Okla., Norman. {SC succession}

66. **Booth, W. E.** 1941. Algae as pioneers in plant succession and their importance in erosion control. Ecology 22: 38-47. {algae soil}

67. **Booth, W. E.** 1941. Revegetation of abandoned fields in Kansas and Oklahoma. Amer. J. Bot. 28: 415-422. {NC Logan Canadian Cleveland Pottawatomie Lincoln Payne succession}

68. **Borchert, J. E.** 1950. The climate of the central North American Grassland. Ann. Assoc. Amer. Geogr. 60: 1-39. {climatology}

69. **Bowers, K. A. W.** 1975. The pollination ecology of *Solanum rostratum* (Solanaceae). Amer. J. Bot. 62: 633-638. {SC Marshall flora}

70. **Bowles, M. L.** 1983. The tallgrass prairie orchids *Platanthera leucophaea* (Nutt.) Lindl. and *Cypripedium candidum* Muhl. ex Willd. and implications toward management. Natural Areas J. 3(4): 14-37. {flora}

71. **Boyd, J. W., D. S. Murray, and R. J. Tyrl.** 1984. Silverleaf nightshade (*Solanum eleagnifolium*) origin, distribution, and relation to man. Econ. Bot. 38: 210-217. {flora}

72. **Brabander, J. J., and R. E. Masters.** 1985. Bottomland hardwoods of eastern Oklahoma: A special study of their status, trends, and values. U.S. Fish and Wildlife Serv., Oklahoma City and Okla. Dept. Wildlife Conservation, Tulsa. {NE SE wildlife habitat floodplain forest}

73. **Brandenburg, D. M.** 1985. Systematic studies in the Poaceae and Cyperaceae. Ph.D. diss., Univ. Okla., Norman. {flora}

74. **Braun, E. L.** 1950. Deciduous forests of eastern North America. Blakiston, Philadelphia. {tree}

75. **Brown, D. D.** 1968. The possible ecological significance of inhibition of *Euphorbia supina*. M.S. thesis, Univ. Okla., Norman. {allelopathy}

76. **Brown, M. J.** 1931. Comparative studies of the animal communities of oak-hickory forests in Missouri and Oklahoma. Univ. Okla. Biol. Surv. Pub. 3: 225-261. {SC habitat wildlife}

77. **Bruner, J. L.** 1976. Analysis of morphological variation in *Schizachyrium scoparium* (little bluestem) in Oklahoma. M.S. thesis, Okla. State Univ., Stillwater. {grassland prairie flora}

78. **Bruner, W. E.** 1926. Overgrazing from an ecological point of view. Proc. Okla. Acad. Sci. 6: 34-38. {SC grassland prairie grazing}

79. **Bruner, W. E.** 1931. The vegetation of Oklahoma. Ecol. Monogr. 1: 99-188. {forest tree grassland prairie all}

80. **Brunken, J. M.** 1971. Cytological and morphological variation in *Panicum virgatum* L. M.S. thesis, Univ. Okla., Norman. {flora}

81. **Bryan, D. E., and W. A. Drew.** 1961. A partial inventory of insect populations in tallgrass prairie pastures in north central Oklahoma. Proc. Okla. Acad. Sci. 42: 143-158. {NC grassland}

82. **Bryant, T.** 1952. Microclimates of three grassland plots in central Oklahoma. M.S. thesis, Univ. Okla., Norman. {SC grassland prairie}

83. **Buck, P.** 1959. A survey of the tall grass prairies in northeastern Oklahoma. M.S. thesis, Univ. Tulsa, Tulsa. {NE grassland}

84. **Buck, P.** 1962. Relationships of the woody vegetation of the Wichita Mountains Wildlife Refuge to geologic formations and soil types. Ph.D. diss., Univ. Okla., Norman. {Comanche SW forest}

85. **Buck, P.** 1964. Relationships of the woody vegetation of the Wichita Mountains Wildlife Refuge to geological formations and soil types. Ecology 45: 336-344. {SW Comanche forest tree shrub}

86. **Buck, P.** 1983. Distribution and identification of woody plants of Oklahoma in the winter condition. Okla. Acad. Sci., Oklahoma City. {flora all}

87. **Buck, P., and E. H. Levetin.** 1982. Weather patterns and ragweed production in Tulsa, Oklahoma. Ann. Allerg. 49: 272-275. {NE allergy climate}

88. **Buck, P., and E. H. Levetin.** 1982. Ragweed pollen and weather patterns in Tulsa, Oklahoma. Bull. Ecol. Soc. Amer. 63: 171. {NE allergy climate}

89. **Buck, P., and E. H. Levetin.** 1985. An assessment of weather patterns and ragweed pollen in Tulsa over a five year period. Seventh Conf. Biometeorology and Aerobiology, Scottsdale, AZ. {NE allergy climate}

90. **Buck, P., and J. J. Crockett.** 1962. Some additions to Eskew's list of flowering plants of the Wichita Mountains Wildlife Refuge. Proc. Okla. Acad. Sci. 43: 41-42. {SW Comanche flora forest grassland}

91. **Buck, P., and R. W. Kelting.** 1962. A survey of the tall-grass prairie in northeastern Oklahoma. Southw. Natur. 7:163-175. {NE grassland}

92. **Bull, R. Z.** 1932. Vascular plants of Greer County, Oklahoma. M.S. thesis, Univ. Okla., Norman. {SW flora}

93. **Burns, P. Y.** 1941. Ecological studies in an eastern Oklahoma floodplain. Proc. Okla. Acad. Sci. 21: 49-52. {bottomland NE Tulsa forest}

94. **Burrows, G. E., A. J. Pollard, R. J. Tyrl, and W. C. Edwards.** 1985. Toxic plants of Oklahoma: Species causing phytodermatoses. J. Okla. Vet. Med. Assoc. 37: 23-26. {poison}

95. **Burrows, G. E., R. J. Tyrl, and W. C. Edwards.** 1981. Toxic plants of Oklahoma: Locoweeds. J. Okla. Vet. Med. Assoc. 33: 13-15. {poison}

96. **Burrows, G. E., R. J. Tyrl, and W. C. Edwards.** 1981. Toxic plants of Oklahoma: Oak, black locust, Kentucky coffee-tree, and buckeye. J. Okla. Vet. Med. Assoc. 33: 37-40. {poison}

97. **Burrows, G. E., R. J. Tyrl, and W. C. Edwards.** 1981. Toxic plants of Oklahoma: Thornapples and nightshades. J. Okla. Vet. Med. Assoc. 33: 106-109. {poison}

98. **Burrows, G. E., R. J. Tyrl, and W. C. Edwards.** 1981. Toxic plants of Oklahoma: Bitterweeds and sneezeweeds. J. Okla. Vet. Med. Assoc. 33: 73-75. {poison}

99. **Burrows, G. E., R. J. Tyrl, and W. C. Edwards.** 1982. Toxic plants of Oklahoma: Ornamental houseplants. J. Okla. Vet. Med. Assoc. 34: 15-18. {poison}

100. **Burrows, G. E., R. J. Tyrl, and W. C. Edwards.** 1982. Toxic plants of Oklahoma: Hemlocks. J. Okla. Vet. Med. Assoc. 34: 30-32. {poison}

101. **Burrows, G. E., R. J. Tyrl, and W. C. Edwards.** 1982. Toxic plants of Oklahoma: Ornamental shrubs. J. Okla. Vet. Med. Assoc. 34: 62-67. {poison}

102. **Burrows, G. E., R. J. Tyrl, and W. C. Edwards.** 1982. Toxic plants of Oklahoma: Coffeweeds and sennas. J. Okla. Vet. Med. Assoc. 34: 101-105. {poison}

103. **Burrows, G. E., R. J. Tyrl, and W. C. Edwards.** 1983. Toxic plants of Oklahoma: Photosensitizing plants and vetch. J. Okla. Vet. Med. Assoc. 35: 39-43. {poison}

104. **Burrows, G. E., R. J. Tyrl, and W. C. Edwards.** 1983. Toxic plants of Oklahoma: Groundsels and rattlebox. J. Okla. Vet. Med. Assoc. 35: 115-118. {poison}

105. **Burrows, G. E., R. J. Tyrl, and W. C. Edwards.** 1983. Ornamental plants potentially hazardous to cattle. Bovine Pract. 18: 188-194. {poison}

106. **Burrows, G. E., R. J. Tyrl, and W. C. Edwards.** 1983. Toxic plants of Oklahoma: Ferns and horsetails. J. Okla. Vet. Med. Assoc. 35: 2-6. {poison}

107. **Burrows, G. E., R. J. Tyrl, and W. C. Edwards.** 1983. Toxic plants of Oklahoma: Cyanogenic plants. J. Okla. Vet. Med. Assoc. 35: 87-90. {poison}

108. **Burrows, G. E., R. J. Tyrl, and W. C. Edwards.** 1984. Toxic plants of Oklahoma: Milkweeds. J. Okla. Vet. Med. Assoc. 36: 74-79. {poison}

109. **Burrows, G. E., R. J. Tyrl, and W. C. Edwards.** 1984. Toxic plants of Oklahoma: Species causing reproductive disorders. J. Okla. Vet. Med. Assoc. 36: 114-116. {poison}

110. **Burrows, G. E., R. J. Tyrl, and W. C. Edwards.** 1985. Toxic plants of Oklahoma: Species causing acute respiratory distress syndrome. J. Okla. Vet. Med. Assoc. 37: 90-92. {poison}

111. **Burrows, G. E., R. J. Tyrl, and W. C. Edwards.** 1985. Toxic plants of Oklahoma: Cocklebur. J. Okla. Vet. Med. Assoc. 37: 63-64. {poison}

112. **Bush, B. F.** 1930. Ferns of Oklahoma. Amer. Midl. Natur. 12: 91-112. {fern flora}

113. **Butler, D. R.** 1980. The effect of salinity on the distribution of algae at the Salt Plains National Wildlife Refuge, Jet, Oklahoma. M.S. thesis, Univ. Okla., Norman. {NW aquatic Alfalfa}

114. **Butler, G. D.** 1878. A list of some of the most interesting species of plants collected in the Indian Territory. Bot. Gaz. 3: 65-68 & 74-78. {all flora}

115. **Butts, K. O.** 1973. Life history and habitat requirements of burrowing owls in western Oklahoma. M.S. thesis, Okla. State Univ., Stillwater. {PH}

116. **Byler, E. B.** 1946. A preliminary survey of the wild flowering plants of Kay County. M.S. thesis, Okla. A & M Coll., Stillwater. {NC flora}

117. **Cantwell, H. D.** 1981. The changing phytosociology of the oak-hickory forest of the Ozark Mountains of Oklahoma. M.S. thesis, Univ. Okla., Norman. {NE}

118. **Carleton, M. A.** 1892. Observations on the native plants of Oklahoma Territory and adjacent districts. Cont. U.S. Natl. Herb. 1: 220-232. {flora all}

119. **Carmes, M. G., and W. T. Penfound.** 1966. Reproduction of woody plants in the absence of grazing. Proc. Okla. Acad. Sci. 47: 7-11. {SC floodplain bottomland forest shrub tree}

120. **Carpenter, E., J. Tarr, and E. Rice.** 1980. Micronutrients in soils of the tall-grass prairie of Oklahoma. Proc. Okla. Acad. Sci. 60: 43-47. {grassland soil}

121. **Carpenter, J. R.** 1934. Seasonal aspects in certain prairie and savannah communities. M.S. thesis, Univ. Okla., Norman. {SC McClain grassland}

122. **Carpenter, J. R.** 1935. Some observations cn the number and size of samples necessary to make an adequate sampling of an ecological prairie community. Proc. Okla. Acad. Sci. 15: 31. {SC grassland method}

123. **Carpenter, J. R.** 1935. Plant distribution in a central Oklahoma prairie. Proc. Okla. Acad. Sci. 15: 32. {SC grassland}

124. **Carpenter, J. R.** 1937. Fluctuations in biotic communities. III. Aspection in a ravine sere in central Oklahoma. Ecology 18: 80-92. {SC succession prairie grassland}

125. **Carpenter, J. R.** 1939. Fluctuations in biotic communities. V. Aspection in mixed-grass prairie in central Oklahoma. Amer. Midl. Natur. 22: 420-435. {SC grassland}

126. **Carpenter, J. R.** 1939. Ecology of the grassland biome. M.S. thesis, Univ. Okla., Norman. {prairie}

127. **Carpenter, J. R.** 1940. The grassland biome. Ecol. Monogr. 10: 617-684. {prairie shortgrass}

128. **Carpenter, J. W.** 1968. Nesting of the mourning dove in northwest Oklahoma. Proc. Okla. Acad. Sci. 49: 163-169. {NW habitat}

129. **Carter, W. A.** 1967. Ecology of the nesting birds of the McCurtain Game Preserve, Oklahoma. Wilson Bull. 79: 259-272. {SE habitat}

130. **Cason, R. G.** 1942. A study of the native grasses of Comanche County, Oklahoma. M.S. thesis, Okla. A & M Coll., Stillwater. {SW grassland prairie}

131. **Chaffin, W.** 1947. A soil improvement program for Oklahoma. Okla. A & M Coll. Circ. No. 412. {management all}

132. **Christensen, J. M.** 1976. Growth of ten species of preferred deer browse plants under various crown densities. M.S. thesis, Okla. State Univ., Stillwater. {forest wildlife habitat}

133. **Clark, M. B.** 1959. A study of the flowering plants of Tulsa County, Oklahoma, exclusive of grasses, sedges, and rushes. M.S. thesis, Univ. Tulsa, Tulsa. {flora NE}

134. **Clark, O. M.** 1928. The arborescent flora of Oklahoma. M.S. thesis, Univ. Wyo., Laramie. {tree forest all}

135. **Clark, T. W.** 1967. The life-form concept and mule deer habitat in Oklahoma. Proc. Okla. Acad. Sci. 48: 23-27. {PH wildlife}

136. **Clement, I. D.** 1948. *Sida* in Oklahoma. Rhodora 50: 99. {flora}

137. **Clingenpeel, J. A.** 1980. Hydrologic and nutrient relationships in a pine-hardwood forest in southeast Oklahoma. M.S. thesis, Okla. State Univ., Stillwater. {SE}

138. **Collins, E. I.** 1977. Vegetation analysis of bobwhite habitat of northeastern Oklahoma rangeland. M.S. thesis, Okla. State Univ., Stillwater. {NE grassland wildlife}

139. **Collins, S. L.** 1987. Interaction of disturbances in tallgrass prairie: a field experiment. Ecology 68: 1243-1250. {NC Canadian fire grazing grassland diversity disturbance}

140. **Collins, S. L., P. G. Risser, and E. L. Rice.** 1981. Ordination and classification of mature bottomland forests in northcentral Oklahoma. Bull. Torrey Bot. Club. 108: 152-165. {NC cluster phytosociology}

141. **Collins, S. L., and D. E. Adams.** 1983. Succession in grasslands: thirty-two years of change in a central Oklahoma tallgrass prairie. Vegetatio 51: 181-190. {SC phytosociology succession}

142. **Collins, S. L., and G. E. Uno.** 1983. The effect of early spring burning on vegetation in buffalo wallows. Bull. Torrey Bot. Club 110: 474-481. {SW Comanche grassland prairie}

143. **Collins, S. L., and G. E. Uno.** 1985. Seed predation, seed dispersal, and disturbance in grasslands: a comment. Amer. Natur. 125: 866-872. {}

144. **Collins, S. L., and S. C. Barber.** 1985. Effects of disturbance on diversity in mixed-grass prairie. Vegetatio 54: 87-94. {SW Comanche fire grassland buffalo bison prairie dog}

145. **Collins, V. E., and W. T. Penfound.** 1966. The relation of stock pond vegetation to geological formations in Marshall County. Proc. Okla. Acad. Sci. 47: 11-13. {SC aquatic}

146. **Conant, S.** 1972. Vegetation structure in a tallgrass prairie. M.S. thesis, Univ. Okla., Norman. {NC Osage grassland}

147. **Conant, S., and P. G. Risser.** 1974. Canopy structure of a tall-grass prairie. J. Range Manage. 27: 313-318. {NC Osage grassland}

148. **Conner, D., D. J. Huddleston, L. A. Pfiester, and S. Thompson.** 1978. A third species of *Enteromorpha* (a marine Chlorophycean) for Oklahoma. Proc. Okla. Acad. Sci. 58: 110. {algae aquatic}

149. **Cooper, J. M.** 1965. Some algae of Lake Carl Blackwell, Oklahoma, exclusive of diatoms. M.S. thesis, Okla. State Univ., Stillwater. {aquatic NC Payne}

150. **Cooper, J. M.** 1972. Spatial and temporal variation in productivity, species diversity, and pigment diversity of periphyton in a stream receiving domestic and oil refinery effluents. Ph.D. diss., Okla. State Univ., Stillwater. {aquatic algae}

151. **Cooper, J. M.** 1985. Additions to the Oklahoma flora: *Sphenoclea zeylanica* (Campanulaceae) and *Ceratophyllum echinatum* (Ceratophyllaceae). Proc. Okla. Acad. Sci. 65: 67. {flora SE McCurtain}

152. **Coppock, R., C. Ely, R. Ficken, and M. Smith.** 1955. An evaluation of the quadrat method in the blackjack-post oak forest. Proc. Okla. Acad. Sci. 36: 49-50. {SC method}

153. **Coryell, P. L.** 1952. A study of the angiosperms, exclusive of the Gramineae, Cyperaceae, and Juncaceae, native to or naturalized in Payne County, Oklahoma. M.S. thesis, Okla. A & M Coll., Stillwater. {NC flora}

154. **Couch, G. C., and E. L. Rice.** 1948. Vegetative habit and reproduction of *Desmidium grevillii* (Kütz) de Bary. Amer. J. Bot. 35: 482-486. {SE McCurtain aquatic algae}

155. **Couch, R. W., and E. N. Nelson.** 1983. *Myriophyllum* in Oklahoma. Proc. Okla. Acad. Sci. 63: 103-104. {aquatic}

156. **Couch, R., and P. Mace.** 1978. Eurasian water milfoil in Oklahoma. Proc. Okla. Acad. Sci. 58: 4-5. {aquatic}

157. **Coyner, W. R.** 1938. Insect distribution and seasonal succession in overgrazed and normal grassland. M.S. thesis, Univ. Okla., Norman. {NC prairie habitat grazing}

158. **Cozad, G., R. Webb, P. Swartz, and J. Clark.** 1953. The effect of moisture on the experimental determination of the metabolic activity of soil microorganisms. Proc. Okla. Acad. Sci. 34: 119-120. {}

159. **Cramer, G. L.** 1975. Effects of various plant and environmental factors on post oak response to 2,4,5-T. M.S. thesis, Okla. State Univ., Stillwater. {SE forest management}

160. **Cress, M.** 1941. After-ripening and germination of seed of *Juniperus virginiana* L. and *Juniperus scopulorum* Sarg. M.S. thesis, Okla. A & M Coll., Stillwater. {}

161. **Crockett, J. J.** 1962. Preliminary studies in the evaluation of grassland sampling techniques in tall-grass prairie sites. Proc. Okla. Acad. Sci. 43: 43-46. {NC method}

162. **Crockett, J. J.** 1962. Grassland communities of the Wichita Mountains Wildlife Refuge. Ph.D. diss., Univ. Okla., Norman. {SW Comanche prairie}

163. **Crockett, J. J.** 1964. Influence of soils and parent materials on grasslands of the Wichita Mountains Wildlife Refuge. Ecology 45: 326-335. {Comanche SW prairie}

164. **Crockett, J. J.** 1965. Effect of mowing on a relict tall grass prairie. Proc. Okla. Acad. Sci. 46: 1-2. {NC grassland management}

165. **Cross, G. L.** 1936. The distribution of *Ephedra* in Oklahoma. Proc. Okla. Acad. Sci. 17: 65-67. {SW flora}

166. **Cross, H. G., G. W. Stevens, and C. W. Shannon.** 1917. The animal and plant life in Oklahoma. Okla. Geol. Surv. Circ. No. 6. {flora classification all}

167. **Dahlem, M. J.** 1972. Low intensity forest sampling through use of stem frequency distribution and population parameters. M.S. thesis, Okla. State Univ., Stillwater. {method}

168. **Dale, E. E., Jr.** 1946. A preliminary survey of the flora of the Arbuckle Mountains. M.S. thesis, Univ. Okla., Norman. {SC Carter Murray}

169. **Dale, E. E., Jr.** 1956. A preliminary survey of the flora of the Arbuckle Mountains. Texas J. Sci. 8: 41-73. {SC Carter Murray}

170. **Dale, E. E., Jr.** 1959. The grasslands of Platt National Park, Oklahoma. Southw. Natur. 4: 45-60. {SC grassland prairie Murray}

171. **Dale, E. E., Jr.** 1965. Final report on vegetation and microenvironments of Platt National Park. Report to National Park Service. {SC Murray grassland forest}

172. **Dalrymple, R. L.** 1964. Vegetation response following winged elm (*Ulmus alata* Michx.) and oak (*Quercus* spp.) control, winged elm browse and winged elm understory growth. M.S. thesis, Okla. State Univ., Stillwater. {SE management}

173. **Dalrymple, R. L., D. D. Dwyer, and P. W. Santlemann.** 1964. Vegetational responses following winged elm and oak control in Oklahoma. J. Range Manage. 17: 249-252. {SE management}

174. **Daniel, H. A., and H. M. Elwell.** 1943. Investigations in erosion control and reclamation of eroded land at the Red Plains Experiment Station, Guthrie, Okla. U.S.D.A. Tech. Bull. No. 837. {NC soil revegetation management}

175. **Davis, C. A.** 1964. Components of the habitat of the bobwhite quail in Payne County, Oklahoma. Ph.D. diss., Okla. State Univ., Stillwater. {NC wildlife}

176. **Davy, R. H.** 1956. The black-tailed prairie dog (*Cynomys ludovicianus*) as a member of a biotic community. M.S. thesis, Okla. State Univ., Stillwater. {grassland prairie habitat}

177. **DeArment, R. D.** 1950. Evaluation of Payne County, Oklahoma farmlands and vegetation patterns for bobwhite quail. M.S. thesis, Okla. A & M Coll., Stillwater. {NC habitat}

178. DeGruchy, H. B. 1956. Water fluctuations as a factor in the life of six of the higher plants of Oklahoma. Proc. Okla. Acad. Sci. 37: 45-46. {NC aquatic}

179. DeGruchy, J. H. B. 1937. A preliminary study of the larger aquatic plants of Oklahoma with special reference to their value in fish culture. M.S. thesis, Okla. A & M Coll., Stillwater. {management}

180. DeGruchy, J. H. B. 1938. A preliminary study of the larger aquatic plants of Oklahoma with special reference to their value in fish culture. Okla. A & M Coll. Bull. No. 4. {management all}

181. DeGruchy, J. H. B. 1952. Water fluctuation as a factor in the life of the higher plants of a 3300 acre lake in the Permian Red Beds of central Oklahoma. Ph.D. diss., Okla. A & M Coll., Stillwater. {aquatic NC Payne}

182. Dent, T. C. 1969. IOPB chromosome number reports. Taxon 18: 433. {flora}

183. Dent, T. C. 1969. Relationships of two isolated groups of sugar maple in central Oklahoma to eastern and western species. Ph.D. diss., Univ. Okla., Norman. {flora SW NC Comanche Caddo Canadian}

184. Dent, T. C., and R. C. Adams. 1983. Relationships of two isolated groups of sugar maples (*Acer saccharum* Marshall ssp. *saccharum*) in west central Oklahoma to eastern and western species. Rhodora 85: 439-456. {SW NC Canadian Caddo Comanche flora}

185. Derdeyn, C. H. 1975. Manipulating central Oklahoma rangeland for bobwhite quail. M.S. thesis, Okla. State Univ., Stillwater. {NC Osage habitat management}

186. Diehl, S. G. 1953. The vegetation of the Wichita Mountains Wildlife Refuge. M.S. thesis, Okla. A & M Coll., Stillwater. {SW Comanche forest grassland prairie}

187. Doing, J. V. 1973. Plant associations as site predictors in the pine-hardwood tension zones in southeastern Oklahoma. M.S. thesis, Okla. State Univ., Stillwater. {SE forest classification management}

188. Donaldson, D. D. 1965. Brush control and the welfare of lesser prairie chickens in western Oklahoma. Proc. Okla. Acad. Sci. 46: 221-228. {NW management habitat}

189. Donaldson, D. D. 1969. Effects on lesser prairie chickens of brush control in western Oklahoma. Ph.D. diss., Okla. State Univ., Stillwater. {NW management habitat}

190. Donselman, H. M., and H. L. Flint. 1982. Genecology of eastern redbud (*Cercis canadensis*). Ecology 63: 962-971. {forest}

191. Dooley, K. 1983. Description and dynamics of some western oak forests in Oklahoma. Ph.D. diss., Univ. Okla., Norman. {SW ordination cluster phytosociology}

192. Dooley, K. L., and S. L. Collins. 1984. Ordination and classification of western oak forests in Oklahoma. Amer. J. Bot. 71: 1221-1227. {SW cluster phytosociology}

193. **Downing, R. L.** 1957. An evaluation of ground nesting by mourning doves in northwestern Oklahoma. M.S. thesis, Okla. State Univ., Stillwater. {NW PH wildlife habitat}

194. **Duck, L. G., and J. B. Fletcher.** 1943. A game type map of Oklahoma. Okla. Game and Fish Dept., Okla. City. {classification all}

195. **Duck, L. G., and J. B. Fletcher.** 1945. A survey of the game and furbearing animals of Oklahoma. Okla. Game and Fish Comm. Bull. No. 3. {classification habitat all}

196. **Dunn, C. W.** 1980. Identifying the prairie-forest transition using remote sensing. M.S. thesis, Univ. Okla., Norman. {classification}

197. **Dvorak, A. J.** 1973. Some morphological and physiological effects of density and density-water in *Bouteloua curtipendula* and *Bouteloua gracilis*. Ph.D. diss., Univ. Okla., Norman. {grassland}

198. **Dwyer, D. D.** 1958. An annotated plant list for Adams Ranch. M.S. thesis, Ft. Hays Kansas State Coll., Hays. {NC Osage prairie grassland flora}

199. **Dwyer, D. D., and P. W. Santlemann.** 1964. A comparison of post oak-blackjack oak communities in two major soil types in north central Oklahoma. Okla. Agr. Exp. Sta. Bull. B-626. {NC Payne forest}

200. **Dyksterhuis, E. J.** 1957. The savannah concept and its use. Ecology 38: 435-442. {classification forest prairie}

201. **Edin, I. C.** 1985. A taxonomic investigation of the *Polygonum pensylvanicum* L. (Polygonaceae) complex. M.S. thesis, Okla. State Univ., Stillwater. {flora}

202. **Edwards, W. C., G. E. Burrows, and R. J. Tyrl.** 1984. Toxic plants of Oklahoma: Cotton and cottonseed. J. Okla. Vet. Med. Assoc. 36: 14-16. {poison}

203. **Edwards, W. C., G. E. Burrows, and R. J. Tyrl.** 1984. Toxic plants of Oklahoma: Clovers. J. Okla. Vet. Med. Assoc. 36: 30-32. {poison}

204. **Ellison, L.** 1960. Influence of grazing on plant succession of rangelands. Bot. Rev. 26: 1-78. {grassland prairie}

205. **Elsharkawi, E. M.** 1968. Water relations of some grasses with phreatophytic properties. Ph.D. diss., Okla. State Univ., Stillwater. {}

206. **Elwell, H. M.** 1945. Seed hay for regrassing eroded land. Proc. Okla. Acad. Sci. 25: 43-45. {NC reclamation revegetation soil management}

207. **Elwell, H. M.** 1967. Herbicides for release of shortleaf pine and native grass. Weeds 15: 104-107. {SE forest Pinus management}

208. **Elwell, H. M., H. A. Daniel, and M. B. Cox.** 1950. Brush control and pasture development in the Red Plains. Agron. J. 42: 390-394. {NC management}

209. **Elwell, H. M., J. W. Slosser, and H. A. Daniel.** 1939. Revegetation and gully-control experiments in the Red Plains Region. Soil Conservation 5: 17-20. {NW soil management}

210. **Elwell, H. M., and M. B. Cox.** 1950. New methods of brush control for more grass. J. Range Manage. 3: 46-51. {NC grassland management}

211. **Endicott, R. L.** 1971. Bio-economic analysis of even aged shortleaf pine stands in southeastern Oklahoma. M.S. thesis, Okla. State Univ., Stillwater. {SE forest management}

212. **England, C. M., and E. L. Rice.** 1956. A comparison of the soil fungi of a tall-grass prairie and of an abandoned field in central Oklahoma. Bot. Gaz. 118: 186-190. {SC McClain}

213. **England, R. B.** 1958. A quantitative comparison of the soil algae of central Oklahoma prairie and woodland soils. M.S. thesis, Univ. Okla., Norman. {SC forest}

214. **Engle, F. E.** 1934. Survey of a wagon road from Fort Smith to the Colorado River. Chron. Okla. 12: 74-75. {presettlement}

215. **Erteeb, F. B.** 1983. The genetic relationship between *Nama hispidum* and *Nama stevensii*. M.S. thesis, Okla. State Univ., Stillwater. {flora}

216. **Eskew, C. T.** 1937. The flowering plants of the Wichita National Forest. M.S. thesis, Univ. Okla., Norman. {SW Comanche flora}

217. **Eskew, C. T.** 1938. The flowering plants of the Wichita Mountains Wildlife Refuge. Amer. Midl. Natur. 20: 695-703. {SW Comanche flora}

218. **Estes, J. R., R. J. Tyrl, and J. N. Brunken (eds.).** 1981. Grasses and grasslands: systematics and ecology. Univ. Okla. Press, Norman. {prairie}

219. **Estes, J. R., and R. J. Tyrl (eds.).** 1976. The grasses and grasslands of Oklahoma. Status of current research on the native and introduced grasses of the state. Ann. Okla. Acad. Sci. 6: 1-182. {prairie grassland pasture}

220. **Estes, J. R., and R. J. Tyrl.** 1976. Classification of Oklahoma grasses. Ann. Okla. Acad. Sci. 6: 141-161. {prairie grassland flora all}

221. **Estes, J. R., and R. J. Tyrl.** 1981. The generic concept and circumscription in the Triticeae: an end paper. In: Grasses and grasslands: systematics and ecology. (ed. by Estes, J. R., R. J. Tyrl, and J. N. Brunken.) Univ. Okla. Press, Norman. {}

222. **Eyles, D. E., J. L. Robertson, and G. W. Jex.** 1944. A guide and key to the aquatic plants of the southeastern United States. USPHS Pub. Health Bull. No. 286. {SE}

223. **Fancher, G. A.** 1981. Natural variation in rooting ability of western provenances of shortleaf pine. M.S. thesis, Okla. State Univ., Stillwater. {forest}

224. **Farqua, H.** 1975. Interseeding and paraquat effects on central and eastern rangeland vegetation. M.S. thesis, Okla. State Univ., Stillwater. {NC NE grassland management}

225. **Featherly, H. I.** 1931. A new species of grama grass. Bot. Gaz. 91: 103-105. {Comanche SW flora}

226. Featherly, H. I. 1938. Grasses of Oklahoma. Okla. A & M Coll. Tech. Bull. No. 3. {flora all}

227. Featherly, H. I. 1938. A new *Triodia* from Oklahoma. Rhodora 40: 243-244. {flora}

228. Featherly, H. I. 1940. The drowning of trees in Lake Carl Blackwell. Proc. Okla. Acad. Sci. 20: 43-44. {NC Payne flooding}

229. Featherly, H. I. 1941. Silting and forest succession on Deep Fork in southwestern Creek County, Oklahoma. Proc. Okla. Acad. Sci. 21: 63-65. {NC floodplain bottomland forest tree}

230. Featherly, H. I., and C. S. Russell. 1939. The ferns of Oklahoma (revised). Okla. A. & M Exp. Sta. Circ. No. C-80. {flora all}

231. Ferguson, Z. T. 1946. Carrying capacity of pastures in Muskogee County, Oklahoma. M.S. thesis, Okla. A & M Coll., Stillwater. {NE grassland management}

232. Fernald, M. L. 1938. A flora of Oklahoma by T. R. Stemen and W. S. Myers – A review. Rhodora 40: 179-183. {}

233. Fernald, M. L. 1950. Gray's manual of botany, 8th edition. American Book Co., New York. {NE SE NC SC flora}

234. Fisher, C. E. 1950. The mesquite problem in the southwest. J. Range Manage. 3: 60-71. {Prosopis management}

235. Fitch, C. H. 1900. Woodland of Indian Territory. U.S. Geol. Surv. Rep. No. 21: 603-672. {forest}

236. Flowers, M. V. 1940. A monographic study of the genus *Aster* in Oklahoma. M.S. thesis, Univ. Okla., Norman. {flora}

237. Fogg, G. G. 1966. Studies in the cembroid pines. M.S. thesis, Univ. Okla., Norman. {PH flora}

238. Foreman, C. T. 1947. The Cross Timbers. Star Printery, Muskogee, OK. {upland forest tree}

239. Foreman, G. 1937. Adventures on the Red River. A journal of Capt. R. B. Marcy. Univ. Okla. Press, Norman. {SE SC SW presettlement}

240. Foreman, G. 1939. Marcy and the gold seekers. Univ. Okla. Press, Norman. {presettlement}

241. Fowells, H. A. 1965. Silvics of forest trees of the United States. Agr. Handbook No. 271, USDA For. Serv., Washington. {flora tree distribution maps}

242. Frank, W. J. 1950. Rodent populations and their reactions to grazing intensities on sand sagebrush grasslands in the southern Great Plains Region. Ph.D. diss., Okla. A & M Coll., Stillwater. {NW habitat management}

243. Gaertner, F. 1955. Honeylocust (*Gleditsia triacanthos* L.) in shelterbelts of western Oklahoma. M.S. thesis, Okla. A & M Coll., Stillwater. {SW management}

244. Gage, H. A. 1908. A preliminary list of the plants of the Arbuckle Mountains. B.A. thesis, Univ. Okla., Norman. {SC flora Murray Carter}

245. Galloway, L. A. 1963. The vegetation of an eroding canyon in Canadian County, Oklahoma. M.S. thesis, Univ. Okla., Norman. {NW}

246. Galloway, L. A. 1964. The vegetation of an actively eroding canyon in Canadian County, Oklahoma. Proc. Okla. Acad. Sci. 45: 20-23. {NW}

247. Gardner, C., J. Jewell, H. Dunn, and I. Mahmoud. 1957. Effects of mowing on a native tall grass prairie in central Oklahoma. Proc. Okla. Acad. Sci. 38: 30-31. {SC grassland management}

248. Gardner, R. G. 1958. A vegetational analysis of the Phillips Agricultural Demonstration Project Ranch, Foraker, Oklahoma. M.S. thesis, A & M Coll. of Texas, College Station. {NC Osage prairie grassland}

249. Gaskill, A. 1905. Why prairies are treeless. Science N. S. 22: 55-56. {fire burning grassland}

250. Gay, C. W. J. 1964. The effects of nitrogen fertilization on native vegetation under clipping, grazing, and burning. M.S. thesis, Okla. State Univ., Stillwater. {Osage NC prairie grassland N management}

251. Gay, C. W., and D. D. Dwyer. 1965. The effects of one year's nitrogen fertilization on native vegetation under clipping and burning. J. Range Manage. 18: 273-277. {Osage NC N prairie grassland management}

252. Gehlbach, F. R., and R. C. Gardner. 1983. Relationships of sugar maples (*Acer saccharum* and *A. grandidentatum*) in Texas and Oklahoma with special reference to relict populations. Texas J. Sci. 35: 231-237. {flora SW NC}

253. Gentry, J. L., R. J. Tyrl, P. G. Risser, and J. J. Crockett. 1978. Field study of the threatened and endangered plants of southeastern Oklahoma. Okla. Biol. Surv., Norman (report to USDI, Bur. Land Mgt.). {SE flora}

254. Gibson, D. J. 1981. Plant communities and the potential for ecotypic differentiation on lead and zinc mine spoils in northeastern Oklahoma. M.S. thesis, Univ. Okla., Norman. {NE Pb Zn Cd succession}

255. Gibson, D. J. 1982. The natural revegetation of lead/zinc mine spoil in northeastern Oklahoma. Southw. Natur. 27: 425-436. {Pb Zn NE Ottawa reclamation}

256. Gibson, D. J., F. L. Johnson, and P. G. Risser. 1985. Revegetation of unreclaimed coal strip mines in Oklahoma. II. Plant communities. Reclam. Reveg. Res. 4: 31-47. {phytosociology ordination cluster NE SE}

257. Gibson, D. J., and P. G. Risser. 1982. Evidence for the absence of ecotypic development in *Andropogon virginicus* (L.) on metalliferous mine wastes. New. Phytol. 92: 589-599. {NE Ottawa Pb Zn Cd}

258. Goertz, J. W. 1962. The influence of habitat quality on density of cotton rat population. Ph.D. diss., Okla. State Univ., Stillwater. {NC Payne}

259. Goodman, G. J. 1934. *Calopogon pulchellus* from Oklahoma. Rhodora 36: 420-421. {flora}

260. Goodman, G. J. 1934. *Glottidium vesicarium* in Oklahoma. Rhodora 36: 118-119. {flora}

261. Goodman, G. J. 1936. Further notes on Oklahoma plants. Rhodora 38: 239-240. {flora}

262. Goodman, G. J. 1956. A new species of *Streptanthus*. Rhodora 58: 354-355. {flora}

263. Goodman, G. J. 1958. Spring flora of central Oklahoma. Univ. Okla. Dup. Serv., Norman. {SC NC}

264. Goodman, G. J., C. A. Lawson, and J. R. Massey. 1978. The Oklahoma botanical travels of G. W. Stevens. Proc. Okla. Acad. Sci. 58: 144-150. {history flora}

265. Goodman, G. J., and C. A. Lawson. 1978. The botanical itineraries of A. H. Van Vleet. Chron. Okla. 56: 322-330. {flora history}

266. Gould, C. N. 1903. Notes on trees, shrubs, and vines in the Cherokee Nation. Trans. Kans. Acad. Sci. 18: 145-146. {NE flora}

267. Gould, C. N. 1930. Petroleum and surface vegetation. Proc. Okla. Acad. Sci. 10: 110-114. {SC tree forest}

268. Gould, L. 1928. Notes on the phenology of middle western spring flowers. Proc. Okla. Acad. Sci. 8: 60-62. {SC}

269. Gould, L. 1929. Phenological notes for 1928 on middle western spring flowers. Proc. Okla. Acad. Sci. 9: 39-42. {SC phenology}

270. Grace, M. C. 1937. A study of the genus *Psoralea* in Oklahoma. M.S. thesis, Univ. Okla., Norman. {flora}

271. Graves, J. E., and W. E. McMurphy. 1969. Burning and fertilization for range improvement in central Oklahoma. J. Range Manage. 22: 165-168. {NC Payne grassland management}

272. Gray, F., and H. M. Galloway. 1959. Soils of Oklahoma. Okla. State Univ. Misc. Pub. 56, Stillwater. {soil all}

273. Gray, F., and M. H. Roozitalab. 1976. Benchmark and key soils of Oklahoma. Okla. State Univ. Agr. Exp. Sta. Bull. MP-97. {classification}

274. Gregg, J. 1844. Commerce of the prairies: or, The journal of a Santa Fe trader, during eight expeditions across the great western prairies, and a residence of nearly nine years in northern Mexico. (two volumes) H. G. Langley, New York. {forest prairie presettlement}

275. Griffin, D. C., and V. W. Proctor. 1964. A population study of *Chara zeylanica* in Texas, Oklahoma, and New Mexico. Amer. J. Bot. 51: 120-124. {SC aquatic}

276. Grummer, B. D. 1977. A biosystematic study of the relationship of *Nama hispidum* and *Nama stevensii*. M.S. thesis, Okla. State Univ., Stillwater. {flora}

277. **Hake, D. R., J. Powell, J. K. McPherson, P. L. Claypool, and G. L. Dunn.** 1984. Water stress of tallgrass prairie plants in central Oklahoma. J. Range Manage. 37: 147-151. {NC}

278. **Halbert, L. T.** 1970. Economical harvesting systems for scattered stands of cottonwood in central Oklahoma. M.S. thesis, Okla. State Univ., Stillwater. {SC NC forest management}

279. **Hall, G. E.** 1950. A pre-impoundment biological survey of Wister Reservoir, Oklahoma. M.S. thesis, Univ. Okla., Norman. {Le Flore SE}

280. **Hall, M. T.** 1947. A taxonomic study of the genus *Juniperus* in Oklahoma. M.S. thesis, Univ. Okla., Norman. {flora}

281. **Hall, M. T.** 1952. Variation and hybridization in *Juniperus*. Ann. Mo. Bot. Gard. 39: 1-64. {SC Arbuckle juniper forest}

282. **Hall, M. T., and C. J. Carr.** 1964. Differential selection in juniper populations from the Baum Limestone and Trinity Sand of southern Oklahoma. Butler Univ. Bot. Studies 14: 21-40. {SC forest tree}

283. **Hall, S. L.** 1978. A comparative study of the water relations of *Quercus marilandica* and *Quercus velutina* with respect to their geographic distribution. M.S. thesis, Okla. State Univ., Stillwater. {forest}

284. **Hall, S. L., and J. K. McPherson.** 1980. Geographic distribution of two species of oaks in Oklahoma in relation to seasonal water potential and transpiration rates. Southw. Natur. 25: 411-423. {NC forest tree water Quercus upland}

285. **Halloran, A. F.** 1958. The occurrence of persimmon and buckeye in the Wichita Mountains Wildlife Refuge. Proc. Okla. Acad. Sci. 39: 4-5. {SW Comanche}

286. **Halloran, A. F., and J. D. Haley.** 1962. A bibliography of the Wichita Mountains of Oklahoma. Great Plains J. (Spring 1962): 1-26. {SW Comanche}

287. **Hamilton, M. W.** 1980. Potentially poisonous or otherwise harmful higher plants of Oklahoma. Proc. Okla. Acad. Sci. 60: 54-62. {poison}

288. **Hamp, S., W. Sorenson, and E. Rice.** 1972. A comparison of soil microfungi of two grassland areas in central Oklahoma. Proc. Okla. Acad. Sci. 52: 41-44. {SC grassland prairie}

289. **Hanson, W. R.** 1960. Plants for improving bobwhite habitat in northwestern Oklahoma. Okla. State Univ. Biol. Ser. No. 7, Stillwater. {NW management}

290. **Hargis, I. J.** 1931. A nature study outline on trees and shrubs for the first six grades. M.S. thesis, Okla. A & M Coll., Stillwater. {flora}

291. **Harlan, J. R.** 1960. Production characteristics of Oklahoma forages: Native range. Okla. Agr. Exp. Sta. Bull. B-547. {prairie grassland management}

292. **Harper, H. J.** 1934. A preliminary survey of important factors which affect tree development in western Oklahoma. Proc. Okla. Acad. Sci. 15: 73-77. {NW SW forest}

293. **Harper, H. J.** 1938. Effect of silting on tree development in the flood plain of the North Canadian River in Creek County, Oklahoma. Proc. Okla. Acad. Sci. 18: 46-49. {bottomland forest tree NE}

294. **Harper, H. J.** 1940. Relation of climatic conditions, soil characteristics, and tree development in the southern Great Plains region. Proc. Soil Sci. Soc. Amer. 5: 327-335. {climate forest}

295. **Harper, H. J.** 1961. Drought years in central Oklahoma from 1710 to 1959 calculated from annual rings of post oak trees. Proc. Okla. Acad. Sci. 41: 23-29. {SC NC Johnston Payne Pawnee dendrochronology climate}

296. **Harper, H. J., and H. A. Daniel.** 1934. Chemical composition of certain aquatic plants. Bot. Gaz. 96: 186-189. {NC Payne}

297. **Harper, H. J., and L. E. Rose.** 1944. Effect of silt on natural vegetation and drainage in the flood plain of the North Canadian River, Lincoln Co., Okla. Proc. Okla. Acad. Sci. 24: 80-82. {NC bottomland forest}

298. **Harper, H. J., and L. W. Reed.** 1964. Effects of chemical composition of soil on micronutrients and other elements in Oklahoma forage plants. Okla. Agr. Exp. Sta. Proc. Ser. P-486. {prairie grassland}

299. **Harrington, T. E.** 1982. A comparison of forest vegetation on orphan strip mines and adjacent sites in eastern Oklahoma. M.S. thesis, Univ. Okla., Norman. {NE SE coal forest succession revegetation}

300. **Harris, K. D.** 1984. A limited-range provenance test of open-pollinated progeny of *Carya illinoensis*. M.S. thesis, Okla. State Univ., Stillwater. {management}

301. **Harris, W. F.** 1950. The identification of some of the more common native Oklahoma grasses by vegetative characters. M.S. thesis, Okla. A & M Coll., Stillwater. {prairie grassland flora all}

302. **Harrison, T. P.** 1974. A floristic study of the woody vegetation of the North American Cross Timbers. Ph.D. diss., North Texas State Univ., Denton. {flora forest savanna prairie}

303. **Hayhome, B., and L. A. Pfiester.** 1983. Electrophoretic analysis of soluble enzymes in five freshwater dinoflagellate species. Amer. J. Bot. 70: 1165-1172. {algae aquatic}

304. **Hazell, D. B.** 1964. Forage production, vegetative composition, and plant vigor in relation to range conditions. Ph.D. diss., Okla. State Univ., Stillwater. {Osage NC prairie grassland management}

305. **Hazell, D. B.** 1965. The claypan range site in northern Osage County, Oklahoma. J. Range Manage. 18: 94-96. {NC prairie grassland}

306. **Hazell, D. B.** 1967. Effect of grazing intensity on plant composition, vigor, and production. J. Range Manage. 20: 249-254. {Osage NC prairie grassland management}

307. **Hedrick, J.** 1931. Lichens from the state of Oklahoma. Pap. Mich. Acad. Sci. Arts. Lett. 13: 101-110. {flora all}

308. Hefley, H. M. 1927. A preliminary report on the seasonal aspects of six habitats near Norman, Oklahoma. Proc. Okla. Acad. Sci. 6: 24-33. {SC Cleveland phenology}

309. Hefley, H. M. 1937. Ecological studies on the Canadian River floodplain in Cleveland County, Oklahoma. Ecol. Monogr. 7: 345-402. {SC succession bottomland forest}

310. Henson, W. E. 1941. Early botanists of Oklahoma. M.S. thesis, Okla. A & M Coll., Stillwater. {history}

311. Hess, W. J. 1967. A taxonomic study of the subgenus Pterogonum of the genus *Eriogonum* (Polygonaceae). Ph.D. diss., Univ. Okla., Norman. {NW SC flora}

312. High, G., and R. J. Taylor. 1980. Vegetation resources including endangered and threatened species of the McGee Creek Project, Atoka County, Oklahoma. U.S. Dept. Of the Interior, Bureau of Reclamation. {SE}

313. High, G., and R. J. Taylor. 1984. Field investigations, McGee Creek Project, Oklahoma, *Calamovilfa arcuata* and *Callirhoe bushii*. U.S. Dept. of the Interior, Bureau of Reclamation. {SE}

314. High, G., and R. J. Taylor. 1985. An environmental study of the W. C. Austin Irrigation Project of southwestern Oklahoma. U.S. Dept. of the Interior, Bureau of Reclamation. {SW}

315. Hill, P. S. M. 1977. A study of interspecific hybridization in a natural oak population with particular regard to introgression. M.S. thesis, Univ. Tulsa, Tulsa. {NE forest Tulsa}

316. Hill, P. S. M., and P. Buck. 1980. Interspecific hybridization in a natural oak population with particular regard to introgression. Proc. Okla. Acad. Sci. 60: 48-53. {NE forest Tulsa}

317. Hindman, F. G. 1965. Retrogression of native range vegetation in the sandy high plains. M.S. thesis, Okla. State Univ., Stillwater. {PH grassland short grass management}

318. Hirst, M. E. 1978. The effect of water potential upon the geographic distribution of *Carya texana*. M.S. thesis, Okla. State Univ., Stillwater. {forest}

319. Hoffman, M. G. 1930. Geology and petrology of the Wichita Mountains. Okla. Geol. Surv. Bull. No. 36. {SW geology}

320. Hoisington, P. B. 1937. Aspection in a central Oklahoma tension zone area. M.S. thesis, Univ. Okla., Norman. {SC forest prairie}

321. Holt, J. R. 1981. Studies in the dinoflagellate genera, *Peridinium* and *Peridiniopsis*. M.S. thesis, Univ. Okla., Norman. {algae}

322. Holt., J. R., and L. A. Pfiester. 1981. A survey of auxotrophy in five freshwater dinoflagellates (Pyrrhophyta). Amer. J. Phycology 17: 415-416. {aquatic algae}

323. Holzinger, J. M. 1892. List of plants collected by C. S. Sheldon and M. A. Carlton in the Indian Territory in 1891. Contr. U.S. Natl. Herb. 1: 188-201. {flora presettlement}

324. **Honess, C. W.** 1923. Geology of the southern Ouachita Mountains of Oklahoma. Okla. Geol. Surv. Bull. No. 32. {SE forest geology}

325. **Hopkins, M.** 1936. Oklahoma spring flowers in autumn. Proc. Okla. Acad. Sci. 17: 60-62. {SC flora phenology}

326. **Hopkins, M.** 1938. Isolation of plants in the Oklahoma flora. Proc. Okla. Acad. Sci. 18: 17-20. {disjunct}

327. **Hopkins, M.** 1938. Notes from the herbarium of the University of Oklahoma. Rhodora 40: 426-434. {flora}

328. **Hopkins, M.** 1941. The floristic affinities of the Arbuckle Mountains in Oklahoma. Amer. J. Bot. 28 (abstracts): 16. {SC Carter Murray flora}

329. **Hopkins, M.** 1943. Notes from the Bebb Herbarium of the University of Oklahoma II. Rhodora 45: 265-277. {biography flora}

330. **Hopkins, M., and U. T. Waterfall.** 1943. Notes on Oklahoma plants. Rhodora 45: 113-119. {flora}

331. **Huffine, W. H., and W. C. Elder.** 1960. Effect of fertilization on native grass pastures in Oklahoma. J. Range Manage. 13: 34-36. {NC Payne grassland management}

332. **Hulnik, F. W.** 1974. The effect of fire on the growth of *Rhus glabra* L. and *Smilax bona-nox* L. M.S. thesis, Okla. State Univ., Stillwater. {fire burning forest}

333. **Hunt, C. B.** 1974. Natural regions of the United States and Canada. W. H. Freeman Co., San Francisco. {physiography all}

334. **Hutcheson, H. L.** 1963. Developmental growth of four species of range grasses in north-central Oklahoma. M.S. thesis, Okla. State Univ., Stillwater. {NC grassland prairie}

335. **Hutcheson, H. L.** 1965. Vegetation in relation to slope exposure and geology in the Arbuckle Mountains. Ph.D. diss., Univ. Okla., Norman. {SC Murray prairie forest grassland}

336. **Hutchinson, G. P.** 1969. Mechanisms of secondary succession in a tall grass prairie. Ph.D. diss., Okla. State Univ., Stillwater. {Payne NC grassland}

337. **Hutchinson, G., R. Anderson, and J. Crockett.** 1966. Changes in species composition of grassland communities in response to grazing intensity. Proc. Okla. Acad. Sci. 47: 25-26. {NC prairie management}

338. **Irving, W.** 1835. A tour of the prairies. In crayon miscellany. Carey, Lea, and Blanchard. Philadelphia. {grassland presettlement}

339. **Irving, W.** (edited by J. F. McDermott). 1956. A tour on the prairies. Univ. Okla. Press, Norman. {presettlement}

340. **James, E.** 1823. Account of an expedition from Pittsburg to the Rocky Mountains, performed in the years 1819, 1820. Vol. 2. Carey and Lea, Philadelphia. {presettlement}

341. **Jarman, R.** 1968. Eurasian watermilfoil – A new menace to Oklahoma waters. Proc. Okla. Acad. Sci. 49: 171-174. {SC SW aquatic}

342. **Jarman, R.** 1984. The development of aquatic ecoregions in Oklahoma. Ph.D. diss., Univ. Okla., Norman. {classification all}

343. **Jeffs, R. E., and E. L. Little.** 1930. A preliminary list of the ferns and seed plants of Oklahoma. Univ. Okla. Biol. Surv. Bull. No. 2., Norman. {flora all}

344. **Johnson, F. L.** 1970. Some vegetation-environment relationships in the upland forests of Oklahoma. M.S. thesis, Univ. Okla., Norman. {soil tree}

345. **Johnson, F. L.** 1973. Biomass, annual net primary production and dynamics of six mineral elements in a post oak-blackjack forest. Ph.D. diss., Univ. Okla., Norman. {SC Cleveland upland forest tree soil N P K Ca Mg Mn}

346. **Johnson, F. L.** 1984. Vegetational changes in a black willow forest over a 23-year period. Proc. Okla. Acad. Sci. 64: 11-13. {SC Marshall succession}

347. **Johnson, F. L.** 1986. Woody vegetation of southeastern LeFlore County, Oklahoma, in relation to topography. Proc. Okla. Acad. Sci. 66: 1-6. {SE upland forest ordination phytosociology}

348. **Johnson, F. L., D. J. Gibson, and P. G. Risser.** 1982. Revegetation of unreclaimed coal strip-mines in Oklahoma. I. Vegetation structure and soil properties. J. Appl. Ecol. 19: 453-463. {Ca K Fe Mn Cu Zn N P biomass succession NE SE}

349. **Johnson, F. L., and P. G. Risser.** 1972. Some vegetation-environment relationships in the upland forests of Oklahoma. J. Ecol. 60: 655-663. {NC NE SE SC SW NW}

350. **Johnson, F. L., and P. G. Risser.** 1973. Correlation analysis of rainfall and annual ring index of central Oklahoma blackjack and post oak. Amer. J. Bot. 60: 475-478. {Cleveland SC dendrochronology climate upland forest}

351. **Johnson, F. L., and P. G. Risser.** 1974. Biomass, annual net primary production, and dynamics of six mineral elements in a post oak-blackjack forest. Ecology 55: 1246-1258. {Quercus N P K Ca Mg Mn upland SC Cleveland}

352. **Johnson, F. L., and P. G. Risser.** 1975. A quantitative comparison between an oak forest and an oak savannah in central Oklahoma. Southwestern Nat. 20: 75-84. {Quercus fire prairie tall-grass SC Cleveland}

353. **Johnson, J. C.** 1957. Habitat preferences among representative wintering and breeding birds of the central Oklahoma forest-prairie ecotone. Ph.D. diss., Univ. Okla., Norman. {SC wildlife}

354. **Jones, R. E.** 1960. The life-form concept applied to prairie chicken habitat in Oklahoma. Proc. Okla. Acad. Sci. 40: 134-136. {NC PH Osage Beaver classification wildlife}

355. **Jones, R. E.** 1961. The quantitative phenology of two plant communities in Osage County, Oklahoma. Proc. Okla. Acad. Sci. 42: 31-38. {NC}

356. **Jones, R. E.** 1963. Identification and analysis of lesser and greater prairie chicken habitat. J. Wildl. Manage. 27: 757-778. {PH NC Osage Beaver classification wildlife}

357. **Jones, R. E.** 1964. Habitat used by lesser prairie chickens for feeding related to seasonal behavior of plants in Beaver County, Oklahoma. Southw. Natur. 9: 111-117. {PH short grass wildlife}

358. **Jones, R. W.** 1939. Observations on the effects of erosion on the ecology and maintenance of an artificial lake in Logan County, Oklahoma. Proc. Okla. Acad. Sci. 19: 37-39. {NC aquatic}

359. **Kapustka, L. A.** 1975. Acetylene reduction (nitrogen fixation) and old-field succession in central Oklahoma. Ph.D. diss., Univ. Okla., Norman. {allelopathy soil N}

360. **Kapustka, L. A., S. V. Bermudez, T. H. Morrison, and R. D. Wingo.** 1976. Chemical and biological analysis of a hazardous waste disposal site. Proc. Okla. Acad. Sci. 56: 42-48. {SC McClain pollution}

361. **Kapustka, L. A., and E. L. Rice.** 1976. Acetylene reduction (N-fixation) of glucose amended soils from central Oklahoma old-field succession plots. Southw. Natur. 23: 389-396. {soil SC Cleveland}

362. **Kapustka, L. A., and E. L. Rice.** 1976. Acetylene reduction (N-fixation) in soil and old-field succession in central Oklahoma. Soil Biol. Biochem. 8: 497-503. {allelopathy SC Cleveland grassland prairie}

363. **Kapustka, L. A., and E. L. Rice.** 1978. Symbiotic and asymbiotic N_2-fixation in a tall grass prairie. Soil Biol. Biochem. 10: 553-554. {McClain SC grassland}

364. **Kapustka, L. A., and F. L. Moleski.** 1976. Changes in community structure in Oklahoma old field succession. Bot. Gaz. 137: 7-10. {SC}

365. **Keck, D. W.** 1961. The lichens of north central Oklahoma. Ph.D. diss., Okla. State Univ., Stillwater. {NC flora}

366. **Kelting, R. W.** 1948. Changes in vegetation on stock pond dams near Norman, Oklahoma. M.S. thesis, Univ. Okla., Norman. {SC Cleveland}

367. **Kelting, R. W.** 1951. Vegetation and soil conditions of prairie and pasture plots in central Oklahoma. Ph.D. diss., Univ. Okla., Norman. {SC grassland}

368. **Kelting, R. W.** 1952. A comparison of organic carbon, hydrogen ion concentration, and volume weight of some central Oklahoma soils. Proc. Okla. Acad. Sci. 33: 151-156. {SC}

369. **Kelting, R. W.** 1954. Effects of moderate grazing on the composition and plant production of a native tall grass prairie in central Oklahoma. Ecology 35: 200-207. {SC grassland management}

370. **Kelting, R. W.** 1957. Winter burning in central Oklahoma grassland. Ecology 38: 520-522. {SC prairie fire burning management}

371. **Kelting, R. W., and W. T. Penfound.** 1950. The vegetation of stock pond dams in central Oklahoma. Amer. Midl. Natur. 44: 69-75. {SC}

372. **Kelting, R. W., and W. T. Penfound.** 1953. Literature on the vegetation of Oklahoma. Proc. Okla. Acad. Sci. 34: 126-135. {bibliography all}

373. **Kennedy, M. T.** 1982. Analysis of the diatom (Bacillariophyta) communities of the North Canadian River, Oklahoma. M.S. thesis, Univ. Okla., Norman. {algae}

374. **Kennedy, M., and L. A. Pfiester.** 1984. Additions to the diatom (Bacillariophyceae) flora of Oklahoma. Proc. Okla. Acad. Sci. 60: 37-39. {algae aquatic}

375. **Kennedy, R. K.** 1973. An analysis of selected Oklahoma upland forest stands including both overstory and understory components. Ph.D. diss., Univ. Okla., Norman. {tree shrub}

376. **Kessler, E.** 1987. *Carthamus lanatus* L. (Asteraceae: Cynareae) a potentially serious plant pest in Oklahoma. Proc. Okla. Acad. Sci. 67: 39-43. {grassland}

377. **Kessler, E., and M. B. Kirkham.** 1985. Restoration of eroded prairie with digested sludge. Proc. Okla. Acad. Sci. 65: 25-34. {SC McClain grassland soil erosion management}

378. **Kingery, C. E.** 1963. Grazing use checks on the Wichitas. J. Range Manage. 16: 142-145. {SW Comanche grassland}

379. **Kingsbury, P. J.** 1968. Plankton production and species distribution in the limnological provinces of Oklahoma. M.S. thesis, Univ. Okla., Norman. {algae aquatic all}

380. **Klinglesmith, R. E.** 1917. A preliminary survey of the flowering plants of Cleveland County. M.S. thesis, Univ. Okla., Norman. {SC flora}

381. **Klopatek, J. M.** 1978. The energetics of land use in three Oklahoma counties. Ph.D. diss., Univ. Okla., Norman. {NW SE SC}

382. **Klopatek, J. M., and P. G. Risser.** 1977. The primary productivity profile of Oklahoma. Okla. Biol. Survey Tech. Report No. 3., Norman. {production all}

383. **Klopatek, J. M., and P. G. Risser.** 1981. Energy analyses of forestry practices in oak-pine forests and loblolly pine plantations. Forest Sci. 27: 365-376. {SE biomass production McCurtain management}

384. **Klopatek, J. M., and P. G. Risser.** 1982. Energy analysis of Oklahoma rangelands and improved pastures. J. Range Manage. 35: 637-643. {prairie grassland production}

385. **Koch, A. R.** 1975. Diatoms, including salt-water taxa, from southwestern Oklahoma. Proc. Okla. Acad. Sci. 55: 11-13. {SW algae}

386. **Koford, C. B.** 1958. Prairie dogs, whitefaces, and blue grama. Wildl. Monogr. 3: 1-78. {grassland habitat wildlife}

387. **Kohlbacher, G. C.** 1978. The effects of periodic flooding on upland forest succession. M.S. thesis, Univ. Okla., Norman. {}

388. **Kuchler, A. W.** 1964. Potential natural vegetation of the conterminous United States. Amer. Geogr. Soc. Sp. Pub. No. 36, New York. {vegetation map classification all}

389. Kuchler, A. W. 1970. A biogeographical boundary: The Tatschl Line. Trans. Kans. Acad. Sci. 73: 298-301. {grassland classification}

390. Lahman, M. S. 1931. Observations of the flora of Delaware County, Oklahoma. Proc. Okla. Acad. Sci. 11: 32-34. {NE}

391. Landolt, J. C. 1986. Cellular slime molds (Dictyosteliales) of selected plant communities of southern Oklahoma and northern Texas. Southw. Natur. 31: 267-269. {SE SC SW}

392. Larkin, E. B. 1930. Systematic study of the flowering plants in the vicinity of Norman with ecological notes on the principal formations. M.S. thesis, Univ. Okla., Norman. {SC Cleveland flora}

393. Larson, F. 1940. The role of the bison in maintaining the short grass plains. Ecology 21: 113-121. {short grass}

394. Lawson, C. A. 1970. The genus *Tilia* in Oklahoma. Proc. Okla. Acad. Sci. 50: 104-105. {SE NE flora}

395. Lawson, C. A. 1972. *Galium pedemontanum* (Bell.) All. (Rubiaceae) reported from Oklahoma. Sida 5: 42. {flora}

396. Lawson, C. A. 1976. The genus *Galium* (Rubiaceae) in the southeastern United States. Ph.D. diss., Univ. Okla., Norman. {flora}

397. Leake, D. V. 1945. The algae of Crystal Lake, Cleveland County, Oklahoma. Amer. Midl. Natur. 34: 750-768. {SC aquatic}

398. Leuck, E. E. 1980. Biosystematic studies in the *Echinocereus viridiflorus* complex. Ph.D. diss., Univ. Okla., Norman. {flora}

399. Levetin, E. H., and P. Buck. 1980. Hay fever plants in Oklahoma. Ann. Allerg. 45: 26-32. {allergy}

400. Levetin, E. H., and P. Buck. 1983. Non-allergenic native and cultivated plants in Oklahoma: Landscaping without hay fever. Ann. Allerg. 52: 166-171. {allergy}

401. Levetin, E. H., and P. Buck. 1984. Aeroallergens in Tulsa, Oklahoma. Sixth Intl. Palynological Conf., Calgary. {NE allergy}

402. Levetin, E. H., and P. Buck. 1985. Wind patterns and the seasonal occurrence of *Juniperus* pollen in Tulsa. Seventh Conf. Biometeorology and Aerobiology, Scottsdale, AZ. {NE allergy}

403. Lewis, W. H., and G. A. Thompson. 1985. *Panax quinquefolium* (Araliaceae) in southeastern Oklahoma. Rhodora 87: 131. {SE flora}

404. Lipscomb, B. 1978. *Cyperus surinamensis* (Cyperaceae) new to Arkansas, Kansas, and Oklahoma. Sida 7: 307. {flora}

405. Little, E. L., Jr. 1936. Bryophytes and pteridiophytes of some west central Oklahoma canyons. Proc. Okla. Acad. Sci. 16: 47-52. {SW flora}

406. Little, E. L., Jr. 1936. The bryophytes of Muskogee County, Oklahoma. Bryologist 38: 8-16. {NE flora}

407. Little, E. L., Jr. 1936. The liverworts of Oklahoma. Bryologist 39: 25-34. {flora all}

408. Little, E. L., Jr. 1938. Flora of Muskogee County, Oklahoma. Amer. Midl. Natur. 19: 369-389. {NE flora}

409. Little, E. L., Jr. 1938. The vegetation of Muskogee County, Oklahoma. Amer. Midl. Natur. 19: 559-572. {NE phytosociology}

410. Little, E. L., Jr. 1939. The vegetation of the Caddo County Canyons, Oklahoma. Ecology 20: 1-11. {SW forest tree phytosociology}

411. Little, E. L., Jr. 1944. *Acer grandidentatum* in Oklahoma. Rhodora 46: 445-450. {SW Canadian Caddo Comanche tree forest flora}

412. Little, E. L., Jr. 1950. Botanical notes from southeastern Oklahoma. Proc. Okla. Acad. Sci. 31: 52-56. {SE McCurtain flora}

413. Little, E. L., Jr. 1980. Baldcypress (*Taxodium distichum*) in Oklahoma. Proc. Okla. Acad. Sci. 60: 105-107. {SE McCurtain forest}

414. Little, E. L., Jr. (revised by). 1981. Forest trees of Oklahoma, 12th edition. Okla. For. Div., State Dept. Agric., Okla. City. {flora forest all}

415. Little, E. L., Jr., and C. E. Olmstead. 1931. An ecological study of the Southeastern Oklahoma Protective Unit. Unpub. ms. in Univ. Okla. Library. {SE tree forest}

416. Little, E. L., Jr., and C. E. Olmstead. 1935. Trees and shrubs of the Southeastern Oklahoma Protective Unit. Proc. Okla. Acad. Sci. 16: 52-61. {SE forest}

417. Lodhi, M. A. K. 1970. Allelopathic effects of *Celtis laevigata* Willd. as related to patterning of vegetation. Ph.D. diss., Univ. Okla., Norman. {SC allelopathy}

418. Lodhi, M. A. K., and E. L. Rice. 1971. Allelopathic effects of *Celtis laevigata*. Bull. Torrey Bot. Club 98: 83-89. {SC allelopathy}

419. Lofton, S. M. 1976. Nitrogen fixation in selected prairie legumes as related to succession. Ph.D. diss., Okla. State Univ., Stillwater. {grassland N}

420. Long, A. W. 1970. Ecological factors affecting the distribution of woody vegetation near the Arkansas River, Tulsa County, with special reference to the smoke tree (*Cotinus obovatus*). M.S. thesis, Univ. Tulsa, Tulsa. {NE forest phytosociology}

421. Love, D. 1980. The effect of land use on nest sites selected by Mississippi kites. M.S. thesis, Okla. State Univ., Stillwater. {management habitat wildlife}

422. Love, H. S. 1971. The detection and nature of functional underground connections between members of certain woody species of the Oklahoma Cross Timbers. Ph.D. diss., Okla. State Univ., Stillwater. {NC SC forest}

423. Loveland, R. W. 1972. Grass stand establishment on rough, rocky, wooded lands in eastern Oklahoma. M.S. thesis, Okla. State Univ., Stillwater. {NE SE grassland management}

424. Lovell, J. F. 1967. Net production relations of two tree species in southern Oklahoma. Proc. Okla. Acad. Sci. 48: 2-6. {SC tree forest}

425. Lowe, R. L. 1919. Collecting in Oklahoma. Bryologist 22: 21-22. {flora}

426. Luckhardt, R. L., and H. G. Barclay. 1938. A study of the environment and floristic composition of an oak-hickory woodland in northeastern Oklahoma. Proc. Okla. Acad. Sci. 18: 25-32. {NE forest tree phytosociology}

427. Lynch, R. A. 1979. The algae of Crystal Lake, 1938-1978. M.S. thesis, Univ. Okla., Norman. {Cleveland aquatic}

428. Lynch, R. A., and L. A. Pfiester. 1981. Crystal Lake revisited. Proc. Okla. Acad. Sci. 61: 49-57. {SC Cleveland algae aquatic}

429. Magrath, L. K. 1971. Vascular plants new for Oklahoma, Kansas, and Nebraska. Rhodora 73: 300-304. {flora NE Ottawa}

430. Magrath, L. K. 1973. The native orchids of the prairies and plains region of North America. Ph.D. diss., Univ. Kansas, Lawrence. {flora all}

431. Magrath, L. K. 1980. *Listera australis* Lindley (Orchidaceae) new for Oklahoma. Southw. Natur. 25: 275. {flora}

432. Magrath, L. K. 1984. New native orchids for Oklahoma and Kansas. Proc. Okla. Acad. Sci. 64: 43-44. {flora}

433. Magrath, L. K. 1987. *Glaucium flavum* Crantz (Papaveraceae) and *Silybum marianum* (L.) Gaertn. (Asteraceae) in Oklahoma. Sida 12: 243. {flora}

434. Magrath, L. K., and J. Taylor. 1978. Orchids and other new and interesting plants from Oklahoma. Pub. No. 2, Southeastern Okla. State Univ., Durant. {flora}

435. Mahler, W. F. 1960. The genus *Baccharis* in the southwestern states of Oklahoma, Texas, and New Mexico. M.S. thesis, Okla. State Univ., Stillwater. {flora}

436. Mahler, W. F. 1981. Notes on rare Texas and Oklahoma plants. Sida 9: 76-86. {flora}

437. Mallik, M. A. B. 1964. Relation between soil fungi and seed plants in three successional forest communities in Oklahoma. Ph.D. diss., Univ. Okla., Norman. {succession SC Cleveland}

438. Mallik, M. A. B., and E. L. Rice. 1966. Relation between soil fungi and seed plants in three successional forest communities in Oklahoma. Bot. Gaz. 127: 120-127. {SC Cleveland bottomland}

439. Maloney, M. M. 1942. Revegetation of coal stripped land near Henryetta, Oklahoma. Proc. Okla. Acad. Sci. 22: 123-130. {NE Okmulgee reclamation succession}

440. Maloney, M. M. 1943. Seasonal algal flora on the campus of Catholic College, Guthrie, Oklahoma. M.S. thesis, Univ. Okla., Norman. {NC Logan algae}

441. Mapp, L. R. 1971. A numerical taxonomic study of *Monarda* (Labiatae) in southern Oklahoma. M.S. thesis, Univ. Okla., Norman. {SC flora}

442. **Marcy, R. B.** 1850. Report of Capt. R. B. Marcy on the railroad from Fort Smith to Santa Fe. 31C15 Ex. Doc. No. 45. H. R. (577). {presettlement}

443. **Marcy, R. B.** 1853. Exploration of the Red River of Louisiana. Senate Ex. Doc. No. 54, 32nd. Congress. {presettlement}

444. **Martin, R. E., and J. R. Preston.** 1968. The mammals of Harmon County, Oklahoma. Proc. Okla. Acad. Sci. 49: 42-60. {SW mixed grass sand sage mesquite shinnery floodplain habitat}

445. **Massey, J. R.** 1971. Pollination biology of *Polygala alba* Nutt. (Polygalaceae). Ph.D. diss., Univ. Okla., Norman. {McClain Grady Kiowa Harmon Woods SC SW NW flora}

446. **May, E. B., and W. T. Penfound.** 1967. Efficiencies of the point-center quarter and quadrat methods in forest sampling. Proc. Okla. Acad. Sci. 48: 6-10. {SC method}

447. **May, S. W.** 1974. Microbial decomposition of cellulose and native plant litter in a true prairie. Ph.D. diss., Univ. Okla., Norman. {NC Osage grassland}

448. **McCollom, G. E.** 1933. Preliminary list of flowering plants of Payne County, Oklahoma. M.S. thesis, Okla. A & M Coll., Stillwater. {NC flora}

449. **McCollum, T., J. Sullivan, and J. R. Estes.** 1984. Reproductive biology of *Commelina erecta* (Commelinaceae)., pp. 57-66, in N. Horner, ed.: Festschrift for Walter Dalquist. Texas Tech Univ. Press, Lubbock. {flora}

450. **McCoy, D.** 1976. Roadside flowers of Oklahoma. C & J Printing Co., Lawton, OK. {flora all}

451. **McCoy, D.** 1978. Roadside flowers of Oklahoma. Vol. 2. C & J Printing Co., Lawton, Ok. {flora all}

452. **McCoy, D.** 1980. Roadside wild fruits of Oklahoma. Univ. Okla. Press, Norman. {flora}

453. **McCoy, D.** 1981. Roadside trees and shrubs of Oklahoma. Univ. Okla. Press, Norman. {flora}

454. **McCoy, D. A.** 1946. A taxonomic study of the Umbelliferae in Oklahoma. M.S. thesis, Univ. Okla., Norman. {all flora}

455. **McCoy, D. A.** 1953. A taxonomic study of the vascular plants indigenous to, or naturalized in, Pontotoc County, Oklahoma. Ph.D. diss., Okla. A & M Coll., Stillwater. {SC flora}

456. **McCoy, D. A.** 1958. Vascular plants of Pontotoc County, Oklahoma. Amer. Midl. Natur. 59: 371-396. {SC flora}

457. **McCullough, C. Y.** 1959. Populations and range effects of rodents on the sand sagebrush grasslands of western Oklahoma. Ph.D. diss., Okla. State Univ., Stillwater. {NW habitat}

458. **McCullough, R. B.** 1972. Phenotypic variation and relationships of several variables in shortleaf pine in Oklahoma. M.S. thesis, Okla. State Univ., Stillwater. {SE NE forest}

459. McDonald, C. B. 1974. A floristic study of the native or naturalized vegetation of Washington County, Oklahoma. M.S. thesis, Okla. State Univ., Stillwater. {NE flora}

460. McDonald, C. B. 1976. A floristic study of Washington County, Oklahoma. Proc. Okla. Acad. Sci. 56: 53-54. {NE Washington flora}

461. McElroy, C. A. 1953. A comparison of the soil fungi of a virgin prairie and an abandoned field. M.S. thesis, Univ. Okla., Norman. {SC grassland}

462. McElroy, C., W. Jones, and F. Rinehart. 1952. An investigation of the soil microflora of two grassland plots. Proc. Okla. Acad. Sci. 33: 163-168. {SC prairie}

463. McGregor, R. L., L. K. Magrath, and R. R. Weedon. 1972. New and interesting plants from the Great Plains. Rhodora 74: 378-388. {flora}

464. McGregor, R. L., et al. 1977. Atlas of the flora of the Great Plains. Iowa State Univ. Press, Ames. {tallgrass shortgrass}

465. McGregor, R. L., et al. 1986. Flora of the Great Plains. Univ. Kansas Press, Lawrence. {flora tallgrass shortgrass}

466. McIlvane, E. H. 1956. Shinnery oak can be controlled. Proc. Southern Weed Conf. 9: 95-98. {SW NW management}

467. McIlvane, E. H., and C. Armstrong. 1963. Progress in shinnery oak and sand sage control at Woodward. Prog. Rept. 6301, Okla. Aerial App. Conf., Stillwater. {NW management}

468. McIlvane, E. H., and C. G. Armstrong. 1966. A summary of fire and forage research on shinnery oak rangelands. Proc. Tall Timbers Fire Ecol. Conf. 5: 127-129. {NW SW burning management}

469. McIlvane, E. H., and D. A. Savage. 1949. Spraying 2,4-D by airplane on sand sagebrush and other plants of the southern Great Plains. J. Range Manage. 2: 43-52. {NW grassland management}

470. McIlvane, E. H., and D. A. Savage. 1951. Fourteen year summary of range improvement studies. Bull. U.S. Southern Great Plains Field Sta., Woodward, Ok. {grassland prairie NW management}

471. McLeod, K. W. 1971. An autecological study of *Salix nigra*, especially factors that might limit its distribution. M.S. thesis, Okla. State Univ., Stillwater. {bottomland}

472. McLeod, K. W., and J. K. McPherson. 1973. Factors limiting the distribution of *Salix nigra*. Bull. Torrey Bot. Club. 100: 102-110. {floodplain bottomland}

473. McMillan, C. 1959. The role of ecotypic variation in the distribution of the central grassland of North America. Ecol. Monogr. 29: 285-308. {prairie}

474. McMillen, W. N., and Q. Williams. 1944. Range and pasture studies and grazing recommendations for the southern Great Plains. Okla. Panhandle Agr. Exp. Sta. Bull. No. 67. {PH NW SW grassland prairie}

475. McMurtry, W. M. 1945. Data and observations concerning regrassing abandoned cultivated land in Lincoln, Kay, and Garfield Counties, Oklahoma. M.S. thesis, Okla. A & M Coll., Stillwater. {NC succession grassland}

476. McPherson, J. K., and G. L. Thompson. 1972. Competitive and allelopathic suppression of understory by Oklahoma oak forests. Bull. Torrey Bot. Club. 99: 293-300. {NC tree shrub}

477. McPherson, J. K., and R. J. Tyrl. 1980. Field surveys for noxious weeds in Oklahoma, Project 12-16-5-2301. Report to USDA, Animal and Plant Health Insp. Serv. {management}

478. Means, F. H. 1969. Vascular plants of southeastern Oklahoma from the Sans Bois to the Kiamichi Mountains. Ph.D. diss., Okla. State Univ., Stillwater. {SE flora}

479. Mericle, L. W. 1941. The spermatophyta of Custer County, Oklahoma. M.S. thesis, Univ. Okla., Norman. {NW flora}

480. Middleton, J. W. 1984. The Oklahoma Natural Heritage Program 1977-1984: a critical evaluation. M.S. thesis, Univ. Okla., Norman. {NHP all}

481. Milby, T. H. 1977. Literature on the vegetation of Oklahoma, 1964-1975. Proc. Okla. Acad. Sci. 57: 176-184. {bibliography all}

482. Milby, T. H. 1986. Germination studies of "Caddo" sugar maple (*Acer saccharum*) seeds. Seed Sci. & Technol. 14: 61-69. {Caddo Canadian SW NW germination forest}

483. Milby, T. H., and F. L. Johnson. 1987. Germination of downy brome from southern Kansas, central Oklahoma, and north Texas. J. Range Manage. 40: 534-536. {weed Bromus germination wheat winter annual SC}

484. Milby, T. H., and W. T. Penfound. 1965. Additions to the literature on the vegetation of Oklahoma. Proc. Okla. Acad. Sci. 45: 23-33. {bibliography all}

485. Milton, W. H. 1947. The seeding, development, and utilization of native grasses in Kay County, Oklahoma. M.S. thesis, Okla. A & M Coll., Stillwater. {NC grassland}

486. Moleski, F. L. 1976. Condensed tannins in soil: inputs and effects on microbial populations. Ph.D. diss., Univ. Okla., Norman. {SC Cleveland allelopathy}

487. Morrison, J. A., and J. C. Lewis. 1974. Preliminary observations on mourning doves wintering in southwestern Oklahoma. Proc. Okla. Acad. Sci. 54: 25-33. {SW habitat}

488. Mueller, D. M. J. 1964. A phenological comparison of mowed and unmowed tallgrass prairie sites in north central Oklahoma. M.S. thesis, Okla. State Univ., Stillwater. {NC phenology grassland}

489. Mullen, W. C. 1977. Eleven year biologic-economic evaluation of undesirable hardwood sites converted to pine. M.S. thesis, Okla. State Univ., Stillwater. {SE forest management}

490. **Muller, C. H.** 1951. The significance of vegetative reproduction in *Quercus*. Madroño 11: 129-137. {shinnery Oklahoma Texas New Mexico}

491. **Murray, C. L.** 1974. A vegetation analysis of a pimpled prairie in northeastern Oklahoma. M.S. thesis, Univ. Tulsa, Tulsa. {NE grassland phytosociology}

492. **Myers, W. S.** 1929. A preliminary report on the flora of the Wichita Mountains. M.S. thesis, Univ. Okla., Norman. {SW Comanche}

493. **Navarro, J. A.** 1935. Germination of weed seeds with reference to temperatures and storage conditions. M.S. thesis, Okla. A & M Coll., Stillwater. {}

494. **Nease, F. R.** 1948. Range deterioration in south central Oklahoma. M.S. thesis, Univ. Okla., Norman. {SC grassland prairie}

495. **Neill, R. L.** 1970. Possible role of *Ambrosia psilostachya* on patterning and succession in old fields. Ph.D. diss., Univ. Okla., Norman. {SC allelopathy}

496. **Neill, R. L., and E. L. Rice.** 1971. Possible role of *Ambrosia psilostachya* on pattern and succession in old fields. Amer. Midl. Natur. 86: 344-357. {SC allelopathy grassland}

497. **Nelson, C. D.** 1984. Genetic variation in juvenile characteristics of *Populus deltoides* Bartr. from the southwestern portion of its native range. M.S. thesis, Okla. State Univ., Stillwater. {}

498. **Nelson, E. N., and R. J. Tyrl.** 1978. Hybridization and introgression between *Elymus canadensis* and *Elymus virginicus* (Poaceae). Proc. Okla. Acad. Sci. 58: 32-34. {grass}

499. **Nelson, E. N., and R. W. Couch.** 1985. Aquatic plants of Oklahoma: I. Submersed, floating-leaved, and selected emergent macrophytes. Oral Roberts Univ., Tulsa. {flora aquatic all}

500. **Nelson, G. L.** 1979. *Erigeron geiseri* (Compositae) in Oklahoma. Southw. Natur. 24: 386-387. {SW Comanche flora}

501. **Nelson, P. M.** 1937. The ecological basis of ornamental gardening in Oklahoma. M.S. thesis, Univ. Okla., Norman. {}

502. **Nelson, R. L.** 1973. The Gulf Coastal Plain Ultisol and Alfisol ecosystem: surface geology, soils, and plant relationships. M.S. thesis, Okla. State Univ., Stillwater. {SE forest}

503. **Netherland, L.** 1979. The effect of disturbances in tall grass prairie sites on an index of diversity and equitability. Southw. Natur. 24: 267-274. {NC Payne grassland fire burning}

504. **Neville, O.** 1928. Mosses of Tulsa and Osage Counties. Proc. Okla. Acad. Sci. 8: 38-41. {NE NC}

505. **Nickell, G. L.** 1972. The physiological ecology of upland and lowland *Panicum virgatum*. Ph.D. diss., Univ. Okla., Norman. {SC prairie grassland}

506. **Nighswonger, P. F.** 1967. A study of soil compaction in a tall grass prairie. M.S. thesis, Univ. Okla., Norman. {McClain SC}

507. **Nighswonger, P. F.** 1969. A study of viable seeds in a selection of birds' nests. Ph.D. diss., Univ. Okla., Norman. {SC Cleveland}

508. **Nixon, W. M.** 1946. Revegetation of eroded lands in the High Plains of Texas and Oklahoma. Proc. Soil Sci. Soc. Amer. 11: 549-552. {soil reclamation grassland}

509. **Nuttall, T.** 1821. Journal of travels into the Arkansas Territory during the year 1819, with occasional observations on the manners of the aborigines. T. N. Palmer, Philadelphia. {NE SE presettlement}

510. **Nuttall, T.** 1837. Collections towards a flora of the Territory of Arkansas. Trans. Amer. Phil. Soc. N. S. 5: 139-203. {flora}

511. **Nuttall, T.** (ed. by S. Lottinville). 1980. A journal of travels into the Arkansas Territory during the year 1819. Univ. Okla. Press, Norman. {NE SE}

512. **Olive, E. W.** 1895. Observations upon some Oklahoma plants. Bull. Torrey Bot. Club. 22: 390-395. {flora}

513. **Olmstead, C. E., Jr.** 1967. Effects of bacterial inhibitors and certain related compounds on the growth of higher plants. M.S. thesis, Univ. Okla., Norman. {allelopathy}

514. **Olmstead, C. E., Jr., and E. L. Rice.** 1970. Relative effects of known plant inhibitors on species from first two stages of old field succession. Southw. Natur. 15: 165-173. {SC allelopathy grassland}

515. **Omohundro, T. T.** 1941. Some experiments and observations on revegetation of Oklahoma range lands with native grass by seeding. M.S. thesis, Okla. A & M Coll., Stillwater. {NC grassland management}

516. **Ophel, I. L.** 1950. A limnological study of two artificial lakes in Oklahoma. M.S. thesis, Univ. Okla., Norman. {aquatic algae}

517. **Ortenberger, A. I.** 1928. Plant collections representative of some typical plant communities in western Oklahoma. Proc. Okla. Acad. Sci. 8: 49-52. {SW NW PH vegetation}

518. **Ortenberger, A. I.** 1928. Plant collections representative of some typical communities in eastern Oklahoma. Proc. Okla. Acad. Sci. 8: 53-57. {NE SE vegetation}

519. **Ortenberger, A. I., and R. D. Bird.** 1933. The ecology of the western Oklahoma salt plains. Pub. Univ. Okla. Biol. Surv. No. 5: 49-64. {NW vegetation}

520. **Orth, M. L. P.** 1980. Distribution of the grasses (Gramineae) in relation to plate tectonics. M.S. thesis, Okla. State Univ., Stillwater. {flora}

521. **Osborn, B.** 1941. Biotic type mapping of Oklahoma watersheds. Proc. Okla. Acad. Sci. 22: 31-33. {classification all}

522. **Osborn, B.** 1942. Prairie dogs in shinnery (oak scrub) savannah. Ecology 23: 110-115. {NW}

523. **Osborn, B., and P. Allan.** 1949. Vegetation of an abandoned prairie dog town in tall-grass prairie. Ecology 30: 320-332. {SW grassland}

524. **Osterhaus, C. A.** 1974. Forest type conversion on the "Cross Timbers" area in Oklahoma. M.S. thesis, Okla. State Univ., Stillwater. {NC SC forest}

525. **Overmire, T. G.** 1963. The effect of grazing on habitat utilization of the dickcissel (*Spiza americana*) and Bell's vireo (*Vireo bellii*) in north central Oklahoma. Ph.D. diss., Okla. State Univ., Stillwater. {NC grassland habitat}

526. **Paka, S.** 1969. Relationship of soil properties to site index of shortleaf pine and distribution of tree species in the coastal plain soils of southeastern Oklahoma. Ph.D. diss., Okla. State Univ., Stillwater. {SE forest}

527. **Palmer, E. J.** 1924. The ligneous flora of Rich Mountain, Arkansas and Oklahoma. J. Arnold Arb. 5: 108-134. {LeFlore SE forest tree}

528. **Palmer, E. J.** 1927. On Nuttall's trail through Arkansas. J. Arnold Arb. 8: 24-52. {SE NE}

529. **Palmer, E. J.** 1932. Leaves from a collector's notebook. J. Arnold Arb. 13:417-437. {flora}

530. **Palmer, E. J.** 1934. Notes on some plants of Oklahoma. J. Arnold Arb. 15: 127-134. {flora}

531. **Pancholy, S. K., and E. L. Rice.** 1973. Soil enzymes in relation to old field succession: amylase, cellulase, invertase, dehydrogenase, and urease. Soil Sci. Soc. Amer. Proc. 37: 47-50. {soil}

532. **Pancholy, S., E. Rice, and J. Turner.** 1975. Soil factors preventing revegetation of a denuded area near an abandoned zinc mine in Oklahoma. J. Appl. Ecol. 12: 337-342. {NE Zn soil}

533. **Parenti, R. L.** 1968. Inhibitional effects of *Digitaria sanguinalis* and possible role in old-field succession. Ph.D. diss., Univ. Okla., Norman. {SC allelopathy}

534. **Parenti, R. L., and E. L. Rice.** 1969. Inhibitional effects of *Digitaria sanguinalis* and possible role in old-field succession. Bull. Torrey Bot. Club 96: 70-78. {allelopathy}

535. **Park, W. J.** 1938. Ecological study of western Payne County, Oklahoma land utilization project. M.S. thesis, Okla. A & M Coll., Stillwater. {prairie grassland NC}

536. **Parker, G. R.** 1967. Edaphic and topographic effects on forest communities in Payne County, Oklahoma. M.S. thesis, Okla. State Univ., Stillwater. {NC soil upland bottomland}

537. **Parker, H. L.** 1952. A comparison of the insects of a virgin prairie, a grazed pasture, and abandoned plowed land. M.S. thesis, Univ. Okla., Norman. {SC McClain grassland Oklahoma habitat}

538. **Parks, J. M.** 1968. Effects of certain plants of old-field succession on the growth of blue-green algae. Ph.D. diss., Univ. Okla., Norman. {SC allelopathy}

539. **Parks, J. M., and H. G. Barclay.** 1965. The increasing importance of vines in southern Oklahoma. Proc. Okla. Acad. Sci. 46: 9-16. {SC vines}

540. **Parton, W. J., and P. G. Risser.** 1979. Simulated impact of management practices upon the tallgrass prairie. In: French, N. R. (ed) Perspectives in grassland ecology. Springer-Verlag. p. 135-155. {grassland NC Osage}

541. **Penfound, W. T.** 1948. An analysis of an elm-ash floodplain community near Norman, Oklahoma. Proc. Okla. Acad. Sci. 28: 59. {Cleveland SC bottomland forest Ulmus Fraxinus phytosociology}

542. **Penfound, W. T.** 1952. An outline for ecological life histories of herbaceous vascular hydrophytes. Ecology 33: 122-128. {aquatic}

543. **Penfound, W. T.** 1953. Plant communities of Oklahoma lakes. Ecology 34: 561-583. {aquatic}

544. **Penfound, W. T.** 1956. The phenology of woody plants in Norman, Oklahoma during the spring seasons of 1954, 1955, and 1956. Proc. Okla. Acad. Sci. 37: 60-64. {Cleveland SC}

545. **Penfound, W. T.** 1956. Primary production of vascular aquatic plants. Limnol. and Oceanog. 1: 92-101. {}

546. **Penfound, W. T.** 1956. Phenology of spring weeds in Norman, Oklahoma. Southw. Natur. 1: 121-124. {SC Cleveland}

547. **Penfound, W. T.** 1961. The composition of a black willow forest in south central Oklahoma. Proc. Okla. Acad. Sci. 41: 30-31. {SC Marshall Texoma floodplain bottomland}

548. **Penfound, W. T.** 1962. The savanna concept in Oklahoma. Ecology 43: 774-775. {all classification}

549. **Penfound, W. T.** 1963. The composition of a post oak forest in south-central Oklahoma. Southw. Natur. 8: 114-115. {SC upland}

550. **Penfound, W. T.** 1963. A modification of the point-centered quarter method for grassland analysis. Ecology 44: 175-176. {SC prairie method}

551. **Penfound, W. T.** 1964. The relation of grazing to plant succession in the tall grass prairie. J. Range Manage. 17: 256-260. {SC McClain grassland}

552. **Penfound, W. T.** 1964. Effects of denudation on the productivity of grassland. Ecology 45: 838-845. {production biomass SC prairie management}

553. **Penfound, W. T.** 1964. Vegetational changes in a black willow forest over a four year period. Proc. Okla. Acad. Sci. 45: 39. {SC Marshall Salix willow Texoma succession}

554. **Penfound, W. T.** 1968. Influence of a wildfire in the Wichita Mountains Wildlife Refuge. Ecology 49: 1003-1006. {fire burning SW Comanche prairie savannah}

555. Penfound, W. T., J. S. Shed, and M. C. Jennison. 1964. A plant community dominated by vines. Proc. Okla. Acad. Sci. 45: 41-43. {SC vines}

556. Penfound, W. T., M. C. Jennison, and J. S. Shed. 1964. Replacement of a population of Johnson grass by a vine-forb community. Proc. Okla. Acad. Sci. 45: 40-41. {SC grassland succession Sorghum}

557. Penfound, W. T., and E. L. Rice. 1957. Effects of fencing and plowing on plant succession in a revegetating field. J. Range Manage. 10: 21-22. {SC McClain management}

558. Penfound, W. T., and E. L. Rice. 1957. Plant population changes in a native prairie plot plowed annually over a five-year period. Ecology 38: 148-150. {SC grassland management}

559. Penfound, W. T., and E. L. Rice. 1957. An evaluation of the arms-length rectangle method in forest sampling. Ecology 38: 660-661. {method}

560. Penfound, W. T., and R. W. Kelting. 1950. Some effects of winter burning on a moderately grazed pasture. Ecology 31: 554-560. {SC prairie grassland fire}

561. Perino, C. H. 1971. Systematics of *Chaetopappa* series Effusae. M.S. thesis, Univ. Okla., Norman. {flora}

562. Perino, J. V. 1971. Some aspects of structure and function in Oklahoma old-field succession. M.S. thesis, Univ. Okla., Norman. {grassland phytosociology McClain Pottawatomie SC}

563. Perino, J. V., and P. G. Risser. 1972. Some aspects of structure and function in Oklahoma old-field succession. Bull. Torrey Bot. Club 99: 233-239. {SC Pottawatomie McClain grassland}

564. Perkins, G., J. R. Estes, and R. W. Thorp. 1975. Pollination of *Cnidoscolus texanus* (Euphorbiaceae) in south central Oklahoma. Southw. Natur. 20: 391-396. {SC Marshall old-field}

565. Petranka, J. W. 1977. A study of the role of *Rhus copallina* in the community dynamics of the forest-prairie ecotone in north-central Oklahoma. M.S. thesis, Okla. State Univ., Stillwater. {NC succession phytosociology}

566. Petranka, J. W., and J. K. McPherson. 1979. The role of *Rhus copallina* in the dynamics of the forest-prairie ecotone in north-central Oklahoma. Ecology 60: 956-965. {NC phytosociology succession}

567. Petranka, J. W., and R. Holland. 1980. A quantitative analysis of bottomland communities in south-central Oklahoma. Southw. Natur. 25: 207-214. {SC bottomland floodplain forest tree}

568. Pfiester, L. A. 1977. *Coronastrum*: a rare find. Trans. Amer. Micros. Soc. 96: 163. {aquatic algae}

569. Pfiester, L. A., O. P. Felkner, and W. O. Felkner. 1976. *Enteromorpha*, a marine alga in Oklahoma. Proc. Okla. Acad. Sci. 56: 66. {algae NW Blaine}

570. **Pfiester, L. A., R. Lynch, and T. L. Wright.** 1980. Species composition and diversity of phytoplankton in the Grand River Dam area, Oklahoma. Proc. Okla. Acad. Sci. 60: 63-68. {NE Mayes algae}

571. **Pfiester, L. A., R. Lynch, and T. L. Wright.** 1979. Species composition and diversity of periphyton in the Grand River Dam area, Mayes County, Oklahoma. Southw. Natur. 24: 149-164. {NE algae}

572. **Pfiester, L. A., and S. Terry.** 1978. Additions to the algae of Oklahoma. Southw. Natur. 23: 85-94. {}

573. **Phillipi, A., and R. J. Tyrl.** 1979. The reproductive biology of *Proboscidea louisianica* (Martyniaceae). Rhodora 81: 345-361. {SC flora}

574. **Phillipi, M. A.** 1977. The reproductive biology of *Proboscidea louisianica* (Martynaceae). M.S. thesis, Okla. State Univ., Stillwater. {flora}

575. **Phillips, P.** 1936. The distribution of rodents in overgrazed and normal grasslands in central Oklahoma. Ecology 17: 673-679. {SC prairie habitat}

576. **Phillips, P. G.** 1935. Rodent distribution in overgrazed and normal grasslands. M.S. thesis, Univ. Okla., Norman. {SC prairie grassland grazing habitat}

577. **Pireh, W.** 1978. Cytogeography and reproductive biology of *Achillea millefolium* in Oklahoma and adjacent states. M.S. thesis, Okla. State Univ., Stillwater. {flora}

578. **Pireh, W., and R. J. Tyrl.** 1980. Cytogeography of *Achillea millefolium* in Oklahoma and adjacent states. Rhodora 82: 361-367. {flora}

579. **Polley, H. W.** 1983. An analysis of the vegetation of buffalo wallows in relation to environmental gradients. M.S. thesis, Univ. Okla., Norman. {SC SW ordination prairie}

580. **Polley, H. W., and L. L. Wallace.** 1986. The relationship of plant species heterogeneity to soil variation in buffalo wallows. Southw. Nat. 31: 493-501. {SW Comanche ordination prairie grassland disturbance}

581. **Polley, H. W., and S. L. Collins.** 1984. Relationships of vegetation and environment in buffalo wallows. Amer. Midl. Natur. 112: 178-186. {SC SW prairie grassland ordination}

582. **Porter, C. L.** 1963. A comparative study of upland and bottomland types of *Panicum virgatum* L. (switch grass) in the vicinity of Norman, Ok. Ph.D. diss., Univ. Okla., Norman. {SC Cleveland McClain ecotype}

583. **Porter, C. L.** 1966. An analysis of variation between upland and lowland switchgrass, *Panicum virgatum* L., in central Oklahoma. Ecology 47: 980-992. {SC prairie grassland Cleveland}

584. **Posey, C. E.** 1967. Natural regeneration of loblolly pine 230 miles northwest of its normal range. J. For. 65: 732. {Pinus}

585. **Pound, R., and F. E. Clements.** 1898. The vegetation regions of the prairie province. Bot. Gaz. 25: 381-394. {grassland classification all}

586. **Powell, J., and D. P. Lowry.** 1980. Oak (*Quercus* spp.) sprout growth rates on a central Oklahoma shallow savannah range site. J. Range Manage. 33: 312-313. {NC Payne prairie grassland}

587. **Preston, J. L.** 1942. Establishment and management of tame and native grass pastures in Oklahoma. M.S. thesis, Okla. A & M College, Stillwater. {grassland prairie}

588. **Prier, C. W.** 1923. A systematic and ecological study of the Gramineae, Cyperaceae, and Juncaceae of the vicinity of Norman, Oklahoma. M.S. thesis, Univ. Okla., Norman. {SC Cleveland McClain flora}

589. **Rasmussen, J. A.** 1970. Allelopathic effects of *Sporobolus pyramidatus* on vegetational patterning. Ph.D. diss., Univ. Okla., Norman. {SC SW Cleveland Comanche allelopathy}

590. **Rasmussen, J. A., and E. L. Rice.** 1971. Allelopathic effects of *Sporobolus pyramidatus* on vegetational patterning. Amer. Midl. Nat. 82:309-326. {allelopathy SC SW Oklahoma Cleveland Comanche}

591. **Ray, H. E.** 1971. A biosystematic study of *Artemisia caruthii*. Ph.D. diss., Univ. Okla., Norman. {flora}

592. **Ray, R. J.** 1957. A phytosociological analysis of the tall-grass prairie in northeastern Oklahoma. M.S. thesis, Univ. Tulsa, Tulsa. {NE grassland phytosociology}

593. **Ray, R. J.** 1959. A phytosociological analysis of the tall-grass prairie in northeastern Oklahoma. Ecology 40: 255-261. {NE grassland}

594. **Reeb, J. E.** 1979. Nutrient analysis of selected forbs on clearcut areas in southeastern Oklahoma. M.S. thesis, Okla. State Univ., Stillwater. {SE forest management nutrient}

595. **Reeds, C. A.** 1927. The Arbuckle Mountains, Oklahoma. Okla. Geol. Surv. Circ. No. 14. {SC Murray Carter geology}

596. **Reisen, W. K.** 1976. The ecology of Honey Creek: Temporal patterns of the travertine periphyton and selected physico-chemical parameters, and *Myriophyllum* community productivity. Proc. Okla. Acad. Sci. 56: 69-74. {Murray SC algae}

597. **Rhodes, L. K.** 1980. Correlation between vegetation and geologic formations in Oklahoma. Oklahoma Geology Notes 40: 47-62. {all geology phytosociology soil}

598. **Rice, E. L.** 1949. Growth and development of five species of range grasses in central Oklahoma. Bot. Gaz. 111: 361-377. {SC McClain prairie phenology}

599. **Rice, E. L.** 1952. Phytosociological analysis of a tall-grass prairie in Marshall County, Oklahoma. Ecology 33:112-116. {SC grassland}

600. **Rice, E. L.** 1960. The microclimate of a relict stand of sugar maple in Devil's Canyon in Canadian County, Oklahoma. Ecology 41: 445-453. {NW tree forest}

601. **Rice, E. L.** 1962. The microclimate of sugar maple stands in Oklahoma. Ecology 43: 19-25. {Canadian Muskogee NW NE upland forest tree Acer}

602. **Rice, E. L.** 1963. Vegetation of Beaver's Bend State Park. Okla. Geol. Surv. Guide Book 9: 39-45. {SE McCurtain forest}

603. **Rice, E. L.** 1964. Inhibition of nitrogen-fixing and nitrifying bacteria by seed plants. Ecology 45: 824-837. {succession NC SC allelopathy}

604. **Rice, E. L.** 1965. Inhibition of nitrogen-fixing and nitrifying bacteria by seed plants. II. Characterization and identification of inhibitors. Physiologia Plantarum 18: 255-268. {allelopathy}

605. **Rice, E. L.** 1965. Bottomland forests of north-central Oklahoma. Ecology 46: 708-714. {NC floodplain tree shrub N P}

606. **Rice, E. L.** 1967. A statistical method for determining quadrat size and adequacy of sampling. Ecology 48: 1047-1049. {method phytosociology}

607. **Rice, E. L.** 1968. Inhibition of nodulation of inoculated legumes by pioneer plant species from abandoned fields. Bull. Torrey Bot. Club 95: 346-358. {succession grassland NC SC allelopathy}

608. **Rice, E. L.** 1976. Allelopathy and grassland improvement. In: Estes, J. R., and R. J. Tyrl (eds). The grasses and grasslands of Oklahoma. Ann. Okla. Acad. Sci. No. 6. Noble Foundation, Ardmore. {prairie allelopathy management}

609. **Rice, E. L.** 1977. Some roles of allelopathic compounds in plant communities. Biochemical Systematics and Ecology 5: 201-206. {allelopathy succession nitrification}

610. **Rice, E. L., R. W. Kelting, and W. T. Penfound.** 1954. Effects of oat straw and native hay mulches on plant composition and edaphic factors of a revegetating abandoned field. Proc. Okla. Acad. Sci. 33: 174-179. {SC soil succession}

611. **Rice, E. L., W. T. Penfound, and L. M. Rorbaugh.** 1960. Seed dispersal and mineral nutrition in succession in abandoned fields in central Oklahoma. Ecology 41: 224-228. {SC McClain grassland}

612. **Rice, E. L., and M. A. B. Mallik.** 1977. Causes of decreases in residual carbohydrase activity in soil during old-field succession. Ecology 58: 1297-1309. {allelopathy succession grassland}

613. **Rice, E. L., and R. L. Parenti.** 1978. Causes of decreases in productivity in undisturbed tall grass prairie. Amer. J. Bot. 65: 1091-1097. {SC McClain grassland fire burning}

614. **Rice, E. L., and R. W. Kelting.** 1955. The species-area curve. Ecology 36: 7-11. {methods phytosociology}

615. **Rice, E. L., and S. K. Pancholy.** 1972. Inhibition of nitrification by climax ecosystems. Amer. J. Bot. 59: 1033-1040. {allelopathy SC SE grassland prairie forest}

616. **Rice, E. L., and S. K. Pancholy.** 1973. Inhibition of nitrification by climax ecosystems. II. Additional evidence and possible role of tannins. Amer. J. Bot. 60: 691-702. {allelopathy SC SE grassland prairie forest}

617. Rice, E. L., and S. K. Pancholy. 1974. Inhibition of nitrification by climax ecosystems. III. Inhibitors other than tannins. Amer. J. Bot. 61: 1095-1103. {SC SE allelopathy forest prairie grassland}

618. Rice, E. L., and W. T. Penfound. 1954. Plant succession and yield of living plant material in a plowed prairie in central Oklahoma. Ecology 35: 176-180. {SC grassland production}

619. Rice, E. L., and W. T. Penfound. 1955. An evaluation of the variable-radius and paired-tree methods in the blackjack-post oak forest. Ecology 36: 315-320. {Cleveland SC forest Quercus method}

620. Rice, E. L., and W. T. Penfound. 1956. Composition of a green ash forest near Norman, Oklahoma. Southw. Natur. 1: 145-147. {SC Cleveland bottomland forest}

621. Rice, E. L., and W. T. Penfound. 1959. The upland forests of Oklahoma. Ecology 40: 593-608. {all forest}

622. Rice, M. O. 1940. Abstracts of ecological literature pertaining to Oklahoma. M.S. thesis, Okla. A & M Coll., Stillwater. {all}

623. Risk, P. L. W. 1978. A numerical analysis of the morphological variation in the genus *Aletris* L. in North America. M.S. thesis, Okla. State Univ., Stillwater. {flora}

624. Risser, P. G. 1970. Competitive responses of *Bouteloua curtipendula* (Michx.) Torr. Amer. Midl. Natur. 84: 259-262. {grassland prairie}

625. Risser, P. G. 1970. Preliminary producer data synthesis: 1970 comprehensive network sites. USIBP Grassland Biome Tech. Rep. No.161. Colo. State Univ., Ft. Collins. {NC Osage Oklahoma prairie grassland}

626. Risser, P. G. 1971. Osage site, 1970 report: Primary production. USIBP Grassland Biome Tech. Rpt. 80, Colo. State Univ., Ft. Collins. {NC Osage prairie grassland production biomass}

627. Risser, P. G. 1971. Plant community structure. In: N. R. French (ed) Preliminary analysis of structure and function in grasslands. Range Sci. Dept. Sci. Ser. No. 10, Colo. State Univ., Ft. Collins. {Osage NC prairie grassland}

628. Risser, P. G. 1972. A preliminary compartment model of a tall grass prairie: Osage 1970. USIBP Grassland Biome Tech. Rpt. No. 159. Colo. State Univ., Ft. Collins. {Osage NC Oklahoma prairie grassland}

629. Risser, P. G. 1972. Systems analysis of a tall grass prairie. Proc. Third Midw. Prairie Conf. 3: 48-53. {Osage NC prairie grassland}

630. Risser, P. G. and F. L. Johnson. 1973. Carbon dioxide exchange characteristics of some prairie grass seedlings. Southw. Nat. 18:85-91. {prairie Andropogon Panicum Sorghastrum Bouteloua photosynthesis}

631. Risser, P. G., J. J. Crockett, R. J. Tyrl, J. L. Gentry, and T. A. Zanoni. 1977. A report on the threatened and endangered plant species in the BLM southeast Oklahoma study area. Okla. Biol. Surv., Norman. (report to USDI, Bur. Land Mgt.). {SE flora}

632. **Risser, P. G., and E. L. Rice.** 1971. Diversity in tree species in Oklahoma upland forests. Ecology 52: 876-880. {phytosociology all}

633. **Risser, P. G., and E. L. Rice.** 1971. Phytosociological analysis of Oklahoma upland forest species. Ecology 52: 940-945. {tree shrub all}

634. **Risser, P. G., and R. K. Kennedy.** 1972. Herbage dynamics of a tall grass prairie: Osage, 1971. USIBP Grassland Biome Tech. Rep. No. 173, Colo. St. Univ., Ft. Collins. {Osage NC prairie grassland}

635. **Risser, P. G., and R. K. Kennedy.** 1975. Herbage dynamics of an Oklahoma tallgrass prairie: Osage 1972. USIBP Grassland Biome Tech. Rep. No. 273, Colo. St. Univ., Ft. Collins. {Osage NC prairie grassland}

636. **Risser, P. G., and W. J. Parton.** 1982. Ecosystem analysis of the tallgrass prairie: nitrogen cycle. Ecology 63: 1342-1351. {NE Osage grassland soil N fire burning decomposition}

637. **Risser, P. G., C. E. Girney, H. D. Blocker, S. W. May, W. J. Parton, and J. A. Wiens.** 1981. The true prairie ecosystem. Hutchinson Ross Publ. Co., Stroudsburg, PA. {grassland}

638. **Rochelle, B. P.** 1984. Sediment-precipitation-runoff relationships for the three ephemeral forest watersheds in southeastern Oklahoma. M.S. thesis, Okla. State Univ., Stillwater. {SE hydrology soil}

639. **Rogers, C. M.** 1953. The vegetation of the Mesa de Maya region of Colorado, New Mexico, and Oklahoma. Lloydia 16: 257-290. {PH Cimarron pinyon juniper ponderosa short grass piñon}

640. **Rogers, C. M.** 1954. Some botanical studies in the Black Mesa region of Oklahoma. Rhodora 56: 205-212. {PH Cimarron short grass pinyon piñon juniper ponderosa}

641. **Rollins, R. C.** 1961. *Draba aprica* in Oklahoma. Rhodora 63: 223-225. {flora}

642. **Ross, R. G.** 1980. Studies in the Cactaceae. Ph.D. diss., Univ. Okla., Norman. {flora}

643. **Roux, E. R., and M. Warren.** 1963. Plant succession on abandoned fields in central Oklahoma and in the Transvaal Highveld. Ecology 44: 576-579. {SC grassland}

644. **Runyon, H. E.** 1939. A new sandbur from western Oklahoma: *Cenchrus albertsonii*. Amer. J. Bot. 26: 485-486. {flora}

645. **Rutledge, J. M.** 1907. A list of the flowering plants of Norman and vicinity. B. A. thesis, Univ. Okla., Norman. {SC Cleveland flora}

646. **Rydberg, P. A.** 1931. A short phytogeography of the prairies and Great Plains of central North America. Brittonia 1: 57-66, + plate. {grassland Ozark forest}

647. **Savage, D. A.** 1934. Methods of reestablishing buffalo grass on cultivated land in the Great Plains. U.S.D.A. Circ. No. 328. {grassland succession management}

648. Savage, D. A. 1937. Drought survival of native grass species in the central and southern Great Plains. U.S.D.A. Tech. Bull. No. 549. {prairie grassland climate}

649. Savage, D. A., and D. F. Costello. 1948. The Southern Great Plains. Grass: USDA Yearbook of Agric., p. 503-540. {prairie grassland}

650. Savage, D. A., and H. E. Runyon. 1937. Natural revegetation of abandoned farmland in the central and southern Great Plains. 4th Int. Grassl. Cong., Rep. Sec. 1: 178-183. Aberystwith. {prairie grassland succession}

651. Schaffner, J. F. 1926. Observations on the grasslands of the central United States. Ohio State Univ. Press, Columbus. {prairie}

652. Schemnitz, S. D. 1960. Status of the scaled quail in western Oklahoma. M.S. thesis, Okla. St. Univ., Stillwater. {NW SW}

653. Schemnitz, S. D. 1961. Ecology of the scaled quail in the Oklahoma panhandle. Wildl. Monogr. 8: {grassland PH short grass}

654. Schnell, G. D., F. L. Johnson, and J. L. Gentry. 1979. Flora and fauna of Oklahoma abandoned mine lands. Okla. Biol. Surv., Norman. {NE SE coal wildlife}

655. Schnell, G. D., F. L. Johnson, and P. G. Risser. 1981. Model fish and wildlife plan for a surface coal mine in Oklahoma. Okla. Dept. Mines., Okla. City and Okla. Biol. Surv., Norman {NE}

656. Schnell, G. D., P. G. Risser, and J. F. Helsel. 1977. Factor analysis of tree distribution patterns in Oklahoma. Ecology 58: 1345-1355. {upland forest tree phytosociology}

657. Schnell, G. D., P. G. Risser, and J. F. Helsel. 1981. Shrub distribution patterns in Oklahoma with comparisons to those exhibited by trees. Bull. Torrey Bot. Club 108: 54-66. {forest tree shrub phytosociology}

658. Segelquist, C. A., and R. E. Pennington. 1968. Deer browse in the Ouachita forest in Oklahoma. J. Wildl. Manage. 32: 623-626. {SE habitat}

659. Sellards, E. H., B. C. Tharp, and K. T. Hill. 1923. Investigation of the Red River made in connection with the Oklahoma-Texas boundary suit. Univ. Texas Bull. No. 2327. {SW}

660. Semtner, M. A. 1972. Possible mechanisms of the exclusion of Johnson grass by tall grass prairie. M.S. thesis, Okla. State Univ., Stillwater. {Payne NC prairie grassland Sorghum}

661. Semtner, P. J., R. W. Barker, and J. A. Hair. 1971. The ecology and behavior of the lone star tick (*Acarina*: Ixodidae). II. Activity and survival in different ecological habitats. J. Med. Entomol. 8: 719-724. {NE}

662. Semtner, P. J., and J. A. Hair. 1973. Distribution, seasonal abundance, and hosts of the Gulf Coast tick in Oklahoma. Ann. Entomol. Soc. Amer. 66: 1264-1268. {NE habitat}

663. Semtner, P. J., and J. A. Hair. 1973. The ecology and behavior of the lone star tick. (*Acarina*: Ixodidae). V. Abundance and seasonal dis-

tribution in different habitat types. J. Med. Entomol. 10: 618-628. {NE habitat}

664. Senter, W. R. 1975. Establishment of warm and cool season grass pastures on wooded sites in east central Oklahoma after aerial spraying and burning. M.S. thesis, Okla. State Univ., Stillwater. {Hughes SC grassland management}

665. Seyfer, J. R., and J. Wihlm. 1977. Variation with stream order in species composition, diversity, biomass, and chlorophyll of periphyton in Otter Creek, Oklahoma. Southw. Natur. 22: 455-467. {NC algae}

666. Shannon, C. W. 1913. The trees and shrubs of Oklahoma. Okla. Geol. Surv. Circ. 4: 7-41. {flora all}

667. Shantz, H. L. 1911. Natural vegetation as an indicator of the capabilities of land for crop production in the Great Plains area. USDA BPI Bull. No. 201: 1-100. {}

668. Shantz, H. L. 1923. The natural vegetation of the Great Plains region. Ann. Assoc. Amer. Geog. 13: 81-107. {prairie grassland}

669. Shantz, H. L., and R. Zon. 1924. Natural vegetation. Atlas of American Agriculture, Sec. E: 1-29. {}

670. Sharp, A. J. 1929. Bryophytes of eastern Oklahoma. Proc. Okla. Acad. Sci. 9: 35-38. {NE SE flora moss liverwort}

671. Sharp, A. J. 1929. The bryophytes of Oklahoma: A preliminary study with notes on their ecology and distribution. M.S. thesis, Univ. Okla., Norman. {liverwort moss flora all}

672. Sharp, A. J. 1930. Bryophyta of southeast Oklahoma. I. A preliminary list with notes. Bryologist 33: 45-55. {SE moss liverwort flora}

673. Shed, J. S., and W. T. Penfound. 1963. Distribution of legumes as correlated with surface geology and plant succession. Proc. Okla. Acad. Sci. 44: 2-6. {SC succession legume}

674. Sheedy, J. D. 1971. Calcium, magnesium, phosphorus, and potassium budgets of a tall grass prairie. M.S. thesis, Univ. Okla., Norman. {NC Osage soil prairie grassland Ca Mg P K}

675. Sheedy, J. D., F. L. Johnson, and P. G. Risser. 1973. A model for phosphorus and potassium flux in a tall-grass prairie. Southw. Natur. 18: 135-149. {Osage NC P K soil grassland}

676. Shelford, V. E. 1926. Naturalist's guide to the Americas. Williams and Wilkins Co., Baltimore. {}

677. Sherwood, R. T. B. 1980. Vegetation of the Woods County, Oklahoma sand dunes. Ph.D. diss., Univ. Okla., Norman. {NW}

678. Sherwood, R. T. B., and P. G. Risser. 1979. Dispersion and co-dispersion in Oklahoma upland forests. Bull. Torrey Bot. Club 106: 200-204. {phytosociology}

679. Sherwood, R. T. B., and P. G. Risser. 1980. Annotated checklist of the vascular plants of Little Sahara State Park, Oklahoma. Southw. Natur. 25: 323-338. {Woods NW dune shortgrass sand sage mixed grass}

680. Sheviak, C. J. 1982. Biosystematic study of the *Spiranthes cernua* complex. Bull. 448, New York State Museum, Albany. {flora orchid}

681. Sheviak, C. J., and M. L. Bowles. 1986. The prairie fringed orchids: A pollinator-isolated species pair. Rhodora 88: 267-290. {flora}

682. Shirley, J. C. 1927. A partial list of the angiosperms of Garfield County. Proc. Okla. Acad. Sci. 7: 105-109. {NC flora}

683. Shirley, J. C. 1937. The woody plants of Oklahoma. Ph.D. diss., Univ. Calif., Berkeley. {tree shrub flora all}

684. Shirley, J. C. 1959. Oklahoma forests and woodlands. Proc. Okla. Acad. Sci. 30: 3-10. {all}

685. Siegler, D. S. 1976. Plants of Oklahoma and Texas capable of producing cyanogenic poisons. Proc. Okla. Acad. Sci. 56: 95-100. {poison}

686. Sims, P. L. 1964. The pattern of retrogression of native vegetation in north central Oklahoma. M.S. thesis, Okla. State Univ., Stillwater. {NC prairie grassland}

687. Sims, P. L., J. S. Singh, and W. K. Lauenroth. 1978. The structure and function of ten western North American grasslands. I. Abiotic and vegetational characteristics. J. Ecol. 66: 251-286. {prairie Osage NC biomass production}

688. Sims, P. L., and D. D. Dwyer. 1965. Pattern of retrogression of native vegetation in north central Oklahoma. J. Range Manage. 18: 20-24. {NC Payne grassland}

689. Sims, P. L., and J. S. Singh. 1978. The structure and function of ten western North American grasslands. IV. Compartment transfers and energy flow within the ecosystem. J. Ecol. 66: 983-1010. {Osage NC prairie biomass production}

690. Sims, P. L., and J. S. Singh. 1978. The structure and function of ten North American grasslands. III. Net primary production, turnover, and efficiencies of energy capture and water use. J. Ecol. 66: 573-598. {NC Osage prairie biomass production}

691. Sims, P. L., and J. S. Singh. 1978. The structure and function of ten western North American grasslands. II. Intra-seasonal dynamics in primary producer compartments. J. Ecol. 66: 547-572. {NC Osage prairie biomass production}

692. Singh, G. 1959. Some effects of differential clipping on six native grasses and on one introduced species. M.S. thesis, Okla. State Univ., Stillwater. {NC Payne grass grassland prairie management}

693. Skroch, J. A. 1965. Effect of disturbance on microclimate and vegetation in tall grass prairie. M.S. thesis, Okla. State Univ., Stillwater. {NC Payne prairie grassland microclimate disturbance}

694. Smith, C. C. 1938. Wildlife and roadside erosion in central Oklahoma. Proc. Okla. Acad. Sci. 19: 31-35. {NC SC soil}

695. Smith, C. C. 1939. Biotic and physiographic succession on abandoned eroded farmland. Ph.D. diss., Univ. Okla., Norman. {}

696. Smith, C. C. 1940. The effect of overgrazing and erosion upon the biota of the mixed grass prairie of Oklahoma. Ecology 21: 381-397. {NC SC grassland}

697. Smith, C. C. 1940. Biotic and physiographic succession on abandoned eroded farmland. Ecol. Monogr. 10: 421-484. {NC SC grassland soil}

698. Smith, E. B. 1976. *Hedyotis ouachitana* (Rubiaceae): a new species from the Ouachita Mountains of Arkansas and Oklahoma. Brittonia 28: 453-459. {SE flora}

699. Smith, J. L. 1982. Differences in nitrate reductase activity between species of different stages in old field succession. Ph.D. diss., Univ. Okla., Norman. {allelopathy}

700. Snook, E. C., J. J. Crockett, and R. J. Tyrl. 1975. Herbaceous plants. In: Rare and endangered vertebrates and plants of Oklahoma. USDA Soil Cons. Serv., Stillwater, OK. {flora}

701. Spaulding, G. F. (ed.). 1968. On the western tour with Washington Irving: The journal and letters of Count de Pourtáles. Univ. Okla. Press, Norman. {presettlement}

702. Spess, R. D. 1975. Revegetation of strip mine spoils. M.S. thesis, Okla. State Univ., Stillwater. {SE NE coal reclamation revegetation}

703. Springer, T. L. 1983. Distribution, habitat, and reproductive biology of *Phlox oklahomensis* (Polemoniaceae). M.S. thesis, Okla. State Univ., Stillwater. {flora}

704. Springer, T. L., and R. J. Tyrl. 1984. Chromosome number of *Phlox oklahomensis* Wherry; IOPB Chromosome Reports. Taxon 33: 133. {}

705. Stanford, J. W. 1973. Umaldy T. Waterfall (1910-1971). Rhodora 73: 146-149. {biography}

706. Stanlake, G. J. 1977. A comparison of the predominant soil microflora in eight vegetative communities in Oklahoma. Proc. Okla. Acad. Sci. 57: 86-90. {sand sage shinnery mesquite mixed grass prairie oak-pine oak-hickory}

707. Stebler, A. M., and S. D. Schemnitz. 1955. Habitat description and the life-form concept. Proc. Okla. Acad. Sci. 36: 154-157. {PH wildlife short grass classification}

708. Steffey, O. D. 1950. Some variations of the fruit of *Quercus macrocarpa*. M.S. thesis, Okla. A & M Coll., Stillwater. {bottomland floodplain forest}

709. Stemen, R. R., and W. S. Myers. 1937. Oklahoma flora. Harlow Pub. Co., Okla. City. {all}

710. Stemen, T. R. 1922. A study of the Compositae and Euphorbiaceae of Oklahoma and Cleveland Counties. M.S. thesis, Univ. Okla., Norman. {SC NC flora}

711. Stevens, G. W. 1916. The flora of Oklahoma. unpub. ms. in Harvard Univ. Library, Cambridge, Mass. {all}

712. Stevens, G. W., and C. W. Shannon. 1917. Plant life in Oklahoma. Okla. Geol. Surv. Bull. No. 27: 215-246. {flora all}

713. **Still, K. R.** 1976. Allelopathic inhibition of weed invasion by *Andropogon scoparius* Michx. in climax prairies. Ph.D. diss., Okla. State Univ., Stillwater. {grassland}

714. **Stratton, R.** 1949. Distribution records of Oklahoma flowering plants. Proc. Okla. Acad. Sci. 20: 14-15. {flora}

715. **Strine, J. H.** 1977. Cottonwood site evaluation in Oklahoma. M.S. thesis, Okla. State Univ., Stillwater. {forest management}

716. **Sullivan, J. R.** 1984. Systematic studies in *Physalis* (Solanaceae). Ph.D. diss., Univ. Okla., Norman. {flora}

717. **Svejcar, T., and S. Christiansen.** 1986. The influence of burning on tillering and aboveground biomass partitioning in little bluestem (*Schizachyrium scoparium*). Southw. Natur. 31: 116-118. {prairie grassland fire Canadian NW}

718. **Swartz, P., R. Webb, G. Cozad, and J. Clark.** 1953. A continuation of the investigation of the soil microflora of two grassland plots. Proc. Okla. Acad. Sci. 34: 121-123. {SC prairie}

719. **Tarr, J., G. Botkin, E. L. Rice, E. Carpenter, and M. Hart.** 1980. A broad analysis of fifteen sites in the tall-grass prairie of Oklahoma. Proc. Okla. Acad. Sci. 60: 39-42. {grassland grass}

720. **Tate, R. C.** 1928. Some observations on the spread of mesquite to the north in Cimarron County, Oklahoma. Proc. Okla. Acad. Sci. 8: 58. {PH grassland}

721. **Taylor, C.** 1960. Annotated checklist of the Compositae of Marshall County, Oklahoma. Proc. Okla. Acad. Sci. 40: 39-49. {SC flora}

722. **Taylor, C.** 1961. Ecology and taxonomy of Water Canyon, Canadian County, Oklahoma. M.S. thesis, Univ. Okla., Norman. {NC flora}

723. **Taylor, C. E. S., and R. J. Taylor.** 1963. Comments on the flora of Bryan County. Proc. Okla. Acad. Sci. 44: 6-10. {SC Bryan}

724. **Taylor, C. E. S., and R. J. Taylor.** 1964. *Podophyllum peltatum* f. *deamii* from Bryan County, Oklahoma. Rhodora 66: 167. {SC flora}

725. **Taylor, C. E. S., and R. J. Taylor.** 1983. New species, new combinations, and notes on the goldenrods (*Euthamia* and *Solidago* - Asteraceae). Sida 10: 176-183. {SW flora}

726. **Taylor, C. E. S., and R. J. Taylor.** 1984. *Solidago* (Asteraceae) in Oklahoma and Texas. Sida 10: 223-251. {flora}

727. **Taylor, C. S.** 1976. National Wildlife Refuges – What future? Ecosystems 6: 4-11. {}

728. **Taylor, R. J.** 1965. Shortleaf pine (*Pinus echinata*) in Bryan County, Oklahoma. Southw. Natur. 10: 42-47. {SC Bryan Oklahoma pine Pinus}

729. **Taylor, R. J.** 1967. The relation of forest vegetation to soils and geology in the gulf coastal plain of Oklahoma. Ph.D. diss., Univ. Okla., Norman. {forest Oklahoma SC soil}

730. **Taylor, R. J.** 1977. *Lemnaea fuscina* Borg (Rhodophyta) in Oklahoma. Proc. Okla. Acad. Sci. 57: 166. {SE Pushmataha algae}

731. **Taylor, R. J.** 1984. *Streptanthus squamiformis*. Status report. Dept. of the Army, U.S. Army Corps of Engineers. {SE}

732. **Taylor, R. J.** 1984. *Draba aprica*. Status report. Dept. of the Army, U.S. Army Corps of Engineers. {SE}

733. **Taylor, R. J.** 1984. *Carex latebracteata*. Status report. Dept. of the Army, U.S. Army Corps of Engineers. {SE}

734. **Taylor, R. J.** 1984. *Amorpha ouachitensis*. Status report. Dept. of the Army, U.S. Army Corps of Engineers. {SE}

735. **Taylor, R. J.** 1984. *Calamovilfa arcuata*. Status report. Dept. of the Army, U.S. Army Corps of Engineers. {SE}

736. **Taylor, R. J.** 1984. *Callirhoe bushii*. Status report. Dept. of the Army, U.S. Army Corps of Engineers. {SE}

737. **Taylor, R. J., and C. E. S. Taylor.** 1965. Additions to the vascular flora of Oklahoma. Rhodora 67: 191-193. {}

738. **Taylor, R. J., and C. E. S. Taylor.** 1969. The vascular flora of Oklahoma – additions and comments. Rhodora 71: 215-219. {}

739. **Taylor, R. J., and C. E. S. Taylor.** 1987. Additions to the vascular flora of Oklahoma – IV. Sida 12: 233-237. {}

740. **Taylor, R. J., and C. E. Taylor.** 1978. Additions to the vascular flora of Oklahoma. Sida 7: 361-368. {}

741. **Taylor, R. J., and C. E. Taylor.** 1981. Plants new to Oklahoma, Arkansas, and Texas. Sida 9: 25-28. {flora}

742. **Taylor, R. J., and C. S. Taylor.** 1980. *Veratrum woodii* (*Melanthium woodii*). Status report. Endangered Species Office, U.S. Fish and Wildl. Serv. {SE}

743. **Taylor, R. J., and C. S. Taylor.** 1980. *Amorpha ouachitensis*. Status report. Endangered Species Office, U.S. Fish and Wildl. Serv. {SE}

744. **Taylor, R. J., and C. S. Taylor.** 1980. *Hypoxis longii*. Status report. Endangered Species Office, U.S. Fish and Wildl. Serv. {SE}

745. **Taylor, R. J., and C. S. Taylor.** 1980. *Calamovilfa arcuata*. Status report. Endangered Species Office, U.S. Fish and Wildl. Serv. {SE}

746. **Taylor, R. J., and C. S. Taylor.** 1980. *Monarda stipitatoglandulosa*. Status report. Endangered Species Office, U.S. Fish and Wildl. Serv. {SE}

747. **Taylor, R. J., and C. S. Taylor.** 1980. *Phlox oklahomensis*. Status report. Endangered Species Office, U.S. Fish and Wildl. Serv. {SE}

748. **Taylor, R. J., and C. S. Taylor.** 1980. *Lesquerella angustifolia*. Status report. Endangered Species Office, U.S. Fish and Wildl. Serv. {SE}

749. **Taylor, R. J., and C. S. Taylor.** 1980. *Cuscuta attenuata*. Status report. Endangered Species Office, U.S. Fish and Wildl. Serv. {SE}

750. **Taylor, R. J., and C. S. Taylor.** 1980. *Carex latebracteata*. Status report. Endangered Species Office, U.S. Fish and Wildl. Serv. {SE}

751. Taylor, R. J., and C. S. Taylor. 1980. *Streptanthus squamiformis*. Status report. Endangered Species Office, U.S. Fish and Wildl. Serv. {SE}

752. Taylor, R. J., and C. S. Taylor. 1981. *Phlox pilosa* var. *longipilosa*. Status report. Endangered Species Office, U.S. Fish and Wildl. Serv. {SE}

753. Taylor, R. J., and C. S. Taylor. 1981. Biology and distribution of *Hypoxis longii* Fern. (Amaryllidaceae). Bull. Torrey Bot. Club 108: 482-484. {SE flora}

754. Taylor, R. J., and C. S. Taylor. 1981. *Phlox pilosa* L. var. *longipilosa* – A new combination. Sida 9: 183. {SE flora}

755. Taylor, R. J., and W. T. Penfound. 1961. The grassland communities of the Baum Limestone in Johnston County, Oklahoma. Southw. Natur. 6: 98-99. {SC prairie}

756. Taylor, R., and C. Taylor. 1966. Comments on the distribution of some Oklahoma plants. Proc. Okla. Acad. Sci. 47: 27-28. {flora}

757. Thieret, J. W. 1974. *Allium ampeloprasum* (Liliaceae) and *Trifolium vesiculosum* (Leguminoseae) in Oklahoma. Sida 5: 286-287. {flora}

758. Thieret, J. W. 1976. Floral biology of *Proboscidea louisianica* (Martyniaceae). Rhodora 78: 169-179. {SC}

759. Thieret, J. W. 1977. Juvenile leaves in Oklahoma *Marsilea* (Marsileaceae). Sida 7: 218-219. {flora}

760. Thomas, W. W. 1973. Stem frequency distribution and sampling of the hardwood component of pine-hardwood stands in southeastern Oklahoma. M.S. thesis, Okla. State Univ., Stillwater. {SE forest}

761. Thompson, G. L. 1961. The Chisholm Trail in geographic perspective. Proc. Okla. Acad. Sci. 42: 253-259. {SC NC prairie grassland}

762. Thompson, W. H. 1943. A limnological study of two small artificial lakes. M.S. thesis, Univ. Okla., Norman. {aquatic algae}

763. Todd, R. L. 1933. Fungi of various depths in typical Cleveland County, Oklahoma soils. Proc. Okla. Acad. Sci. 14: 61-64. {SC}

764. Torrey, J. 1828. Some account of a collection of plants made during a journey to and from the Rocky Mountains in the summer of 1820, by Edwin James, M.D., Assistant Surgeon, U.S. Army. Ann. Lyc. Nat. Hist. N. Y. 2: 161-254. {flora presettlement}

765. Troeger, W. W. 1977. Epiphytic diatoms as indicators of water quality in farm and experimental ponds in Oklahoma. M.S. thesis, Univ. Okla., Norman. {algae aquatic}

766. Troeger, W. W. 1978. Survey of algae in farm ponds in Bryan County, Oklahoma. Southw. Natur. 23: 51-62. {SC aquatic}

767. Troeger, W. W. 1978. Epiphytic diatoms in farm ponds and experimental ponds in Bryan County, Oklahoma. Proc. Okla. Acad. Sci. 58: 64-68. {SC algae aquatic}

768. Troeger, W. W. 1981. Epiphytic diatoms as indicators of water quality in farm and experimental ponds in Oklahoma. Proc. Okla. Acad. Sci. 61: 79-80. {algae aquatic}

769. **Troeger, W. W., and A. Olness.** 1980. Plant development and sequence in artificial ponds with and without phosphorus-enriched sediments. Southw. Natur. 25: 411-423. {Bryan SC algae aquatic}

770. **Turner, J. A.** 1974. Loss of ferulic acid from decomposing leaves of *Celtis laevigata* and subsequent isolation of organisms able to use it as the sole carbon source. Ph.D. diss., Univ. Okla., Norman. {SC allelopathy}

771. **Turner, J. A., and E. L. Rice.** 1975. Microbial decomposition of ferulic acid in soil. J. Chem. Ecol. 1: 41-58. {allelopathy}

772. **Tyler, J. D.** 1965. Habitat types used by the bobwhite quail and other wildlife in southwestern Oklahoma. M.S. thesis, Okla. State Univ., Stillwater. {SW wildlife}

773. **Tyler, J. D.** 1970. Vertebrates in a prairie dog town. Proc. Okla. Acad. Sci. 50: 110-113. {SW habitat}

774. **Tyrl, R. J.** 1975. Origin and distribution of polyploid *Achillea* (Compositae) in western North America. Brittonia 27: 135-146. {flora}

775. **Tyrl, R. J.** 1978. Gathering of *Ipomopsis rubra* (Polemoniaceae) nectar by *Bombus* (Apidae). Southw. Natur. 23: 153-154. {}

776. **Tyrl, R. J.** 1980. Identification and mapping of the extant flora at the Deer Creek archaeological site (34Ka 3, Kaw Lake, Oklahoma). Report to U.S. Army, Corps of Engineers. {Kay Osage}

777. **Tyrl, R. J.** 1980. The genus *Achillea*. In: A synonimized checklist of the vascular flora of the United States, Canada, and Greenland (ed. by J. T. Kartesz and R. Kartesz). Univ. North Carolina Press, Chapel Hill. {flora}

778. **Tyrl, R. J.** 1982. Nomenclature of the genus *Achillea*. In: National list of scientific names (ed. by W. Rice, S. Smith, and D. Wasshausen). USDA, Soil. Cons. Serv. Tech. Rept. No. 159., Washington. {flora}

779. **Tyrl, R. J., F. B. Erteeb, J. L. Bruner, W. L. Ezell, J. R. Estes, and B. Grummer.** 1984. A biosystematic study of the relationship of *Nama hispidum* and *Nama stevensii* (Hydrophyllaceae). Southw. Natur. 29: 367-375. {flora}

780. **Tyrl, R. J., J. L. Bruner, and M. A. Phillipi.** 1977. Chromosome numbers in *Nama* and *Proboscidea*; IOPB Chromosome Reports. Taxon 26: 559. {flora}

781. **Tyrl, R. J., J. L. Gentry, P. G. Risser, and J. J. Crockett.** 1979. Field surveys and status evaluation on proposed and candidate endangered plant species in Oklahoma. Okla. Biol. Surv., Norman (report to USDI, Fish and Wildlife Serv.) {flora}

782. **U.S. Army, Corps of Engineers.** 1973. Final environmental statement: Optima Lake, North Canadian River, Oklahoma. Tulsa District, Corps of Engineers. {PH Texas}

783. U.S. Army, Corps of Engineers. 1974. Final environmental statement: Hugo Lake, Kiamichi River, Oklahoma. Tulsa District, Corps of Engineers. {SE Choctaw}

784. U.S. Army, Corps of Engineers. 1975. Final updated and revised environmental statement: Kaw Lake, Arkansas River, Oklahoma. Tulsa District, Corps of Engineers. {NC Kay Osage}

785. U.S. Geological Survey. 1926. Geological map of Oklahoma. USGS, Washington. {geology all}

[442.] **U.S. House of Representatives**. 31st. Congress, 1st. Session. 1850. Route from Fort Smith to Santa Fe ... a report and map of Lt. Simpson ..., also a report on the same subject from Capt. R. B. Marcy ... (House Exec. Doc. 45) Washington ... 89 p., 1 map, 1 plate. (Serial set 577). {presettlement}

[839.] **U.S. Senate.** 23rd. Congress, 2nd. session. 1834. Journal of the expedition of the regiment of dragoons into Indian country in 1834 by Lt. T. B. Wheelock. Senate Doc. 1, No. 1: 73-93. (Serial set 266). {presettlement}

[4.] **U.S. Senate.** 29th. Congress, 1st session. 1846. Journal of Lt. J. W. Abert, from Bent's Fort to St. Louis in 1845. Senate Doc. 8, No. 438: 1-75. (Serial set 477). {presettlement}

[443.] **U.S. Senate.** 32nd. Congress, 2nd. session. 1853. Exploration of the Red River of Louisiana in the year 1852: by Randolph B. Marcy assisted by G. B. McClellan. With reports on the natural history of the country and numerous illustrations. (Sen. Ex. Doc. 54). Washington ... 320 p., numerous plates. (Serial set 666). See also Foreman 1937 (#239). {presettlement}

786. U.S. War Department. 1855. Report of the secretary of War ... (see Chron. Okla. 28: 235-283.) {presettlement}

787. U.S. War Department. 1856. [reports of railroad route expeditions]. 33rd. Cong., 2nd.Sess., Sen. Ex. Doc. 78,[], V.3, pts. I & II. {presettlement}

788. U.S. War Department. 1856. Reports of explorations and surveys ... [long title – deals with railroad route explorations] 33rd. Cong., 2nd. Sess., House Ex. Doc. 91,[], V.4, pt. 5. {presettlement}

789. U.S. War Department. (Lt. A. W. Whipple). 1855. [long title concerning exploration of railroad routes]. 33rd. Congress, 1st. Sess., House Ex. Doc. 129, Vol.2, Part VI. {presettlement}

790. Umber, R. L. 1978. A study of allelopathy in *Helianthus rigidus*. M.S. thesis, Okla. State Univ., Stillwater. {}

791. Ungar, I. A. 1966. Salinity tolerance of marsh plants in Kansas and Oklahoma. Ecology 47: 154-155. {salt NW aquatic}

792. Ungar, I. A. 1968. Species-soil relationships in the Great Salt Plains of northern Oklahoma. Amer. Midl. Natur. 80: 392-406. {soil salt NW}

793. United States Department of Agriculture. 1938. Soils and men. Yearbook of agriculture. U.S. Govt. Printing Office, Washington. {soil all}

794. United States Department of Agriculture. 1941. Climate and man. Yearbook of agriculture. U.S. Govt. Printing Office, Washington. {climate all}

795. United States Department of Agriculture. 1949. Trees. Yearbook of agriculture. U.S. Govt. Printing Office, Washington. {all}

796. University of Tulsa, Faculty of Natural Sciences. 1977. A biological inventory of the Fort Gibson Lake area. U.S. Dept. of the Army, Corps of Engineers, Tulsa Dist. {NE flora}

797. Uno, G. E., and S. L. Collins. 1987. Primary succession on granite outcrops in southwestern Oklahoma. Bull. Torrey Bot. Club 114: 387-392. {SW diversity prairie soil community classification}

798. Uttal, L. J. 1984. *Solidago flaccidifolia* Small (Asteraceae) new to Oklahoma and other goldenrod notes. Sida 10: 324-325. {flora}

799. Van Vleet, A. H. 1902. Plants of Oklahoma. Dept. Geology and Natural History, 2nd. Biennial Rpt., 1901-1902: 138-160. {flora all}

800. Vaughn, J. M. 1981. Systematics of *Aristida dichotoma*, *basiramea*, and *curtisii* (Poaceae). M.S. thesis, Univ. Okla., Norman. {flora}

801. Vowell, J. L. 1980. Descriptive and hydrologic characteristics of three forested watersheds in Oklahoma's Kiamichi Mountains. M.S. thesis, Okla. State Univ., Stillwater. {SE forest}

802. Wallis, C. S. 1953. The spermatophyta of Cherokee County, Oklahoma (exclusive of Gramineae, Cyperaceae, and Juncaceae). M.S. thesis, Okla. A & M Coll., Stillwater. {NE flora}

803. Wallis, C. S. 1957. Additions to the Oklahoma flora from the Oklahoma Ozarks. Proc. Okla. Acad. Sci. 38: 3-5. {NE}

804. Wallis, C. S. 1959. Vascular plants of the Oklahoma Ozarks. Ph.D. diss., Okla. State Univ., Stillwater. {flora NE}

805. Wallis, C. S., and U. T. Waterfall. 1953. Additions to the Oklahoma flora from Cherokee County. Proc. Okla. Acad. Sci. 34: 124-125. {NE}

806. Ward, C. G. 1941. Notes on the growth and fruiting of *Andropogon scoparius*, with some reference to its reaction to a changed environment. M.S. thesis, Okla. A & M Coll., Stillwater. {grassland prairie}

807. Ware, G. H. 1948. A study of revegetation in the floodplain of the South Canadian River. M.S. thesis, Univ. Okla., Norman. {SC Cleveland succession floodplain}

808. Ware, G. H., and W. T. Penfound. 1949. The vegetation of the lower levels of the South Canadian River in central Oklahoma. Ecology 30: 478-484. {SC floodplain bottomland succession}

809. Waterfall, U. T. 1940. Interesting plants of Oklahoma. Rhodora 42: 499-502. {flora}

810. Waterfall, U. T. 1942. The spermatophyta of Oklahoma County, Oklahoma, exclusive of the grasses, sedges, and rushes. M.S. thesis, Univ. Okla., Norman. {NC flora}

811. Waterfall, U. T. 1948. Distributional notes and some minor forms from Oklahoma. Rhodora 50: 91-98. {flora}

812. Waterfall, U. T. 1948. A new species of *Euphorbia* from Oklahoma. Rhodora 50: 63-64. {flora}

813. Waterfall, U. T. 1949. Some results of a summer's botanizing in Oklahoma. Rhodora 51: 18. {flora}

814. Waterfall, U. T. 1949. A new species of *Sarcostemma* from Oklahoma. Rhodora 51: 58-59. {flora}

815. Waterfall, U. T. 1950. Some additions to the Oklahoma flora. Rhodora 52: 19-24, 35-41. {}

816. Waterfall, U. T. 1950. Some results of a third summer's botanizing in Oklahoma. Rhodora 52: 165-175. {flora}

817. Waterfall, U. T. 1951. More additions to the Oklahoma flora. Rhodora 53: 130. {}

818. Waterfall, U. T. 1952. Further studies of the Oklahoma flora. Rhodora 54: 125-131. {}

819. Waterfall, U. T. 1953. The identity of *Hedyotis rosea* Raf. Rhodora 55: 201-203. {NC flora}

820. Waterfall, U. T. 1954. Studies in the composition and distribution of the Oklahoma flora. XXI. Rhodora 56: 157-162. {}

821. Waterfall, U. T. 1954. A new species of *Carex* (Section Phyllostachyae) from Oklahoma. Rhodora 56: 21-23. {flora}

822. Waterfall, U. T. 1960. Changes in status and new combinations for certain taxa in the Oklahoma flora. Rhodora 62: 319-321. {}

823. Waterfall, U. T. 1970. *Monarda stipitatoglandulosa*, a new species from Oklahoma. Rhodora 72: 502-504. {flora}

824. Waterfall, U. T. 1972. Keys to the flora of Oklahoma, 5th edition. Published by the author. {all}

825. Waterfall, U. T., and C. S. Wallis. 1962. Some geographic relationships of the vascular flora of the Oklahoma Ozarks. Proc. Okla. Acad. Sci. 43: 61-63. {NE}

826. Waterfall, U. T., and C. S. Wallis. 1963. A list of the vascular flora of the Oklahoma Ozarks. Proc. Okla. Acad. Sci. 44: 11-22. {NE}

827. Watkins, W. E. 1973. Plant succession following destruction of vegetation by toxins from a zinc roaster and smelter. M.S. thesis, Univ. Okla., Norman. {NE Okmulgee}

828. Watson, L. J. 1975. The geographic variation of *Asclepias tuberosa interior* in Oklahoma. M.S. thesis, Univ. Okla., Norman. {flora}

829. Watson, S. H. 1939. Progress of regrassing in the Southern Great Plains. Soil Cons. 5: 57-59. {grassland}

830. Weaver, J. E. 1954. North American prairie. Johnsen Pub. Co., Lincoln, Nebr. {grassland}

831. Weaver, J. E., and F. W. Albertson. 1956. Grasslands of the Great Plains. Johnsen Pub. Co., Lincoln, Nebr. {prairie}

832. Weaver, J. E., and N. W. Rowland. 1952. Effects of excessive mulch on development, yield, and structure of native grassland. Bot. Gaz. 114: 1-19. {prairie}

833. Webb, W. L. 1950. Biogeographic regions of Texas and Oklahoma. Ecology 31: 427-433. {all classification}

834. Weedon, R. R., and L. K. Magrath. 1973. New and interesting plants from the Great Plains. Southw. Natur. 18: 341-342. {flora}

835. Weese, A. O. 1925. Oklahoma climagraphs and biotic regions. Proc. Okla. Acad. Sci. 5: 91-95. {climate classification all}

836. Weese, A. O. 1939. The effect of overgrazing on insect populations. Proc. Okla. Acad. Sci. 19:95-99. {prairie grassland SW}

837. Welborne, F. F. 1962. The comparative ecology of two canyons and an upland area in west central Oklahoma. Ph.D. diss., Univ. Okla., Norman. {Canadian Caddo SW NC forest}

838. Welch, P. 1929. A reconnaissance of a typical central Oklahoma prairie. M.S. thesis, Univ. Okla., Norman. {SC grassland}

839. Wheelock, T. 1834. Journal of the expedition of the regiment of dragoons into Indian country in 1834. 23rd. Cong., 2nd. sess., Sen. Doc. 1, No. 1: 73-93. (Chron. Ok. 28: 2-41) {bottomland forest presettlement}

[789.] Whipple, A. W. 1855. (see U.S. War Department. 1855). {}

840. White, P. J. 1901. A study of the flora of Oklahoma. M.S. thesis, Univ. Okla., Norman. {all}

841. Whitenton, R. O. 1921. Life zones and zone indicators in Oklahoma. Proc. Okla. Acad. Sci. 1: 43-45. {classification all}

842. Wiedeman, V. E. 1960. Preliminary ecological study of the shinnery oak area of western Oklahoma. M.S. thesis, Univ. Okla., Norman. {NW SW Quercus}

843. Wiedeman, V. E., and W. T. Penfound. 1960. A preliminary study of the shinnery in Oklahoma. Southw. Natur. 5: 117-122. {NW SW}

844. Willemsen, R. W. 1971. Mechanism of seed dormancy in *Ambrosia artemisiifolia*. Ph.D. diss., Univ. Okla., Norman. {germination}

845. Williams, J. E. (1973.) Atlas of the woody plants of Oklahoma. Okla. Biol. Surv., Norman. {tree shrub forest distribution maps all}

846. Williams, S. T., and B. D. Simison. 1927. Washington Irving on the prairie, or, a narrative of a tour in the southwest in the year 1832 by H. L. Ellsworth. American Book Co., Dallas, TX. {}

847. Willson, D. L. 1958. A survey of the soil algae of two vegetation types in central Oklahoma. Ph.D diss., Univ. Okla., Norman. {SC grassland prairie forest}

848. Willson, D. L., and H. S. Forrest. 1957. An exploratory study on soil algae. Ecology 38: 309-313. {SC grassland prairie}

849. Wilson, R. E. 1968. Allelopathy as expressed by *Helianthus annuus*, and its role in old-field succession. Ph.D. diss., Univ. Okla., Norman. {SC Cleveland sunflower allelopathy}

850. Wilson, R. E., and E. L. Rice. 1968. Allelopathy as expressed by *Helianthus annuus* and its role in old-field succession. Bull. Torrey Bot. Club 95: 432-448. {SC sunflower allelopathy}

851. Wilson, W. D. 1968. Shortleaf pine site index – Soils and plant associations on the coastal plain of southeastern Oklahoma. M.S. thesis, Okla. State Univ., Stillwater. {SE Pinus forest management}

852. Wolfe, J. M. 1978. Allelopathic interactions between algae. M.S. thesis, Univ. Okla., Norman. {allelopathy}

853. Woods, E. D. 1985. Environmental parameters and site identification for potential expansion of the pine forests of eastern Oklahoma. M.S. thesis, Okla. State Univ., Stillwater. {NE SE management}

854. Wright, K. L. T. 1984. Studies in the Danthonieae (Poaceae). Ph.D. diss., Univ. Okla., Norman. {flora}

855. Wright, T. L. 1977. *Gloeodendron*: A rare alga in Oklahoma. Proc. Okla. Acad. Sci. 57: 169-170. {SC Cleveland Marshall algae}

856. Wright, T. L. 1977. A comparison of algal succession between farm ponds with sandstone and clay substrates. M.S. thesis, Univ. Okla., Norman. {algae aquatic}

857. Wright, T. L., and L. A. Pfiester. 1978. A comparison of algal composition and seasonal succession between ponds with sandstone and clay substrates. Southw. Natur. 23: 431-456. {algae SC Cleveland aquatic}

858. Zanoni, T. A. 1977. Publications of the Oklahoma Biological Survey. Inf. Bull. No. 1, Okla. Biol. Surv., Norman. {bibliography}

859. Zanoni, T. A., J. L. Gentry, R. J. Tyrl, and P. G. Risser. 1979. Endangered and threatened plants of Oklahoma. Dept. Botany-Micro., Univ. Okla., Norman and Dept. General and Evolutionary Biol., Okla. St. Univ., Stillwater. {rare endangered threatened impact all}

860. Zanoni, T. A., P. G. Risser, and I. H. Butler. 1979. Natural areas for Oklahoma: important areas of natural significance in Oklahoma. Okla. Nat. Heritage Prog., Univ. Okla., Norman. {classification forest grassland aquatic all}

861. Zwick, F. B. 1980. The host plant-insect ecology of *Pyrrhopappus* (Asteraceae: Lactuceae) and *Schinia mitis* (Lepidoptera: Noctuidae). M.S. thesis, Univ. Okla., Norman. {}

862. Zwick, F. B., and J. R. Estes. 1981. Adaptation of the life-history of *Schinia mitis* to its host plant *Pyrrhopappus*. J. Kansas Entomol. Soc. 54: 416-432. {}

Indexes

AUTHOR INDEX
(Numbers refer to Citations)

Abbott, J. E.: 1
Abdul-Wahab, A. S.: 2, 3
Abert, J. W.: 4
Adams, D. B.: 5, 6, 7
Adams, D. E.: 8, 9, 10, 11, 12, 28, 141
Adams, R. C.: 184
Ahshapanek,: D. C. 13, 14, 15
Al Naib, F. A. G.: 16, 17, 18
Al Saadawi, I. S.: 19
Albertson, F. W.: 20, 831
Allameh, A.: 21
Allan, P. F.: 22, 523
Alleman, D. R.: 23
Allgood, F. P.: 24
Allred, B. W.: 25
Amos, B. B.: 26
Anderson, K. L.: 27
Anderson, R. C.: 9, 12, 28, 337
Anonymous: 29
Armstrong, C. G.: 467, 468
Arter, B. S.: 30
Atkins, A. P.: 31
Atwood, J. T.: 32
Baalman, R. J.: 33
Baldock, L. O.: 34
Barber, S. C.: 35, 36, 37, 38, 144
Barclay, H. G.: 39, 40, 41, 426, 539
Barker, B. M.: 42
Barker, R. W.: 661
Barkley, E. A.: 43, 44, 45
Baumgartner, A. M.: 46
Baumgartner, F. M.: 46
Beare, M. H.: 47
Becker, D. A.: 48
Bermudez, S. V.: 360
Bertelson, D. F.: 51, 52
Bessinger, G. H.: 49
Bigham, H. R.: 50
Bird, R. D.: 519
Birdsey, R. A.: 51, 52
Blair, A. P.: 53
Blair, W. F.: 54, 55
Blinn, W. C.: 56
Blocker, H. D.: 637
Blum, U.: 57, 58
Bogue, E. E.: 59, 60
Bokhari, V. G.: 61
Bol, E. L.: 62
Bollinger, C. J.: 63
Boone, N.: 64
Booth, W. E.: 65, 66, 67
Borchert, J. E.: 68
Botkin, G.: 719
Bowers, K. A. W.: 69
Bowles, M. L.: 70, 681
Boyd, J. W.: 71
Brabander, J. J.: 72
Brandenburg, D. M.: 73
Braun, E. L.: 74
Brown, D. D.: 75
Brown, M. J.: 76
Bruner, J. L.: 77, 779, 780
Bruner, W. E.: 78, 79
Brunken, J. M.: 80, 218
Bryan, D. E.: 81
Bryant, T.: 82
Buck, P.: 83, 84, 85, 86, 87, 88, 89, 90, 91, 316, 399, 400, 401, 402
Bull, R. Z.: 92
Burns, P. Y.: 93
Burrows, G. E.: 94, 95, 96, 97, 98, 99, 100, 101, 102, 103, 104, 105, 106, 107, 108, 109, 110, 111, 202
Bush, B. F.: 112
Butler, D. R.: 113
Butler, G. D.: 114
Butler, I. H.: 860
Butts, K. O.: 115
Byler, E. B.: 116
Cantwell, H. D.: 117
Carleton, M. A.: 118
Carmes, M. G.: 119
Carpenter, E.: 120, 719
Carpenter, J. R.: 121, 122, 123, 124, 125, 126, 127
Carpenter, J. W.: 128
Carr, C. J.: 282
Carter, W. A.: 129
Cason, R. G.: 130
Chaffin, W.: 131
Christensen, J. M.: 132
Christiansen, S.: 717
Clark, J.: 158, 718
Clark, M. B.: 133
Clark, O. M.: 134

Clark, T. W.: 135
Claypool, P. L.: 277
Clement, I. D.: 136
Clements, F. E.: 585
Clingenpeel, J. A.: 137
Collins, E. I.: 138
Collins, S. L.: 9, 139, 140, 141, 142, 143, 144, 192, 581, 797
Collins, V. E.: 145
Conant, S.: 146, 147
Conner, D.: 148
Cooper, J. M.: 149, 150, 151
Coppock, R.: 152
Coryell, P. L.: 153
Costello, D. F.: 651
Couch, G. C.: 154
Couch, R. W.: 155, 156, 499
Cox, M. B.: 208, 210
Coyner, W. R.: 157
Cozad, G.: 158, 718
Cramer, G. L.: 159
Cress, M.: 160
Crockett, J. J.: 48, 90, 161, 162, 163, 164, 253, 337, 631, 700, 781
Cross, G. L.: 165
Cross, H. G.: 166
Dahlem, M. J.: 167
Dale, E. E., Jr.: 168, 169, 170, 171
Dalrymple, R. L.: 172, 173
Daniel, H. A.: 174, 208, 209, 296
Davis, C. A.: 175
Davy, R. H.: 176
DeArment, R. D.: 177
DeGruchy, J. H. B.: 178, 179, 180, 181
Dent, T. C.: 182, 183, 184
Derdeyn, C. H.: 185
Diehl, S. G.: 186
Doing, J. V.: 187
Donaldson, D. D.: 188, 189
Donselman, H. M.: 190
Dooley, K.: 191, 192
Downing, R. L.: 193
Drew, W. A.: 81
Duck, L. G.: 194, 195
Dunn, C. W.: 196
Dunn, G. L.: 277
Dunn, H.: 247
Dvorak, A. J.: 197
Dwyer, D. D.: 173, 198, 199, 251, 688
Dyksterhuis, E. J.: 200
Edin, I. C.: 201
Edwards, W. C.: 94, 95, 96, 97, 98, 99, 100, 101, 102, 103, 104, 105, 106, 107, 108, 109, 110, 111, 202
Elder, W. C.: 331
Ellison, L.: 204
Elsharkawi, E. M.: 205
Elwell, H. M.: 174, 206, 207, 208, 209, 210
Ely, C.: 152
Endicott, R. L.: 211
England, C. M.: 212
England, R. B.: 213
Engle, F. E.: 214
Erteeb, F. B.: 215, 779
Eskew, C. T.: 216, 217
Estes, J. R.: 10, 38, 218, 219, 220, 221, 449, 564, 779, 862
Eyles, D. E.: 222
Ezell, W. L.: 779
Fancher, G. A.: 223
Farqua, H.: 224
Featherly, H. I.: 225, 226, 227, 228, 229, 230
Felkner, O. P.: 569
Felkner, W. O.: 569
Ferguson, Z. T.: 231
Fernald, M. L.: 232, 233
Ficken, R.: 152
Fisher, C. E.: 234
Fitch, C. H.: 235
Fletcher, J. B.: 194, 195
Flint, H. L.: 190
Flowers, M. V.: 236
Fly, C. L..: 27
Fogg, G. G.: 237
Foreman, C. T.: 238
Foreman, G.: 239, 240
Forrest, H. S.: 848
Fowells, H. A.: 241
Frank, W. J.: 242
Gaertner, F.: 243
Gage, H. A.: 244
Galloway, H. M.: 272
Galloway, L. A.: 245, 246
Gardner, C.: 247
Gardner, R. C.: 252
Gardner, R. G.: 248
Gaskill, A.: 249
Gay, C. W. J.: 250, 250
Gehlbach, F. R.: 252
Gentry, J. L.: 253, 631, 654, 781, 859
Gibson, D. J.: 254, 255, 256, 257, 348

Girney, C. E.: 637
Goertz, J. W.: 258
Goodman, G. J.: 259, 260, 261, 262, 263, 264, 265
Gould, C. N.: 266, 267
Gould, L.: 268, 269
Grace, M. C.: 270
Graves, J. E.: 271
Gray, F.: 24, 272, 273
Gregg, J.: 274
Griffin, D. C.: 275
Grummer, B. D.: 276, 779
Hair, J. A.: 661, 662, 663
Hake, D. R.: 277
Halbert, L. T.: 278
Haley, J. D.: 286
Hall, G. E.: 279
Hall, M. T.: 280, 281, 282
Hall, S. L.: 283, 284
Halloran, A. F.: 285, 286
Hamilton, M. W.: 287
Hamp, S.: 288
Hanson, W. R.: 289
Hargis, I. J.: 290
Harlan, J. R.: 291
Harper, H. J.: 292, 293, 294, 295, 296, 297, 298
Harrington, T. E.: 299
Harris, K. D.: 300
Harris, W. F.: 301
Harrison, T. P.: 302
Hart, M.: 719
Hayhome, B.: 303
Hazell, D. B.: 304, 305, 306
Hedrick, J.: 307
Hefley, H. M.: 308, 309
Helsel, J. F.: 656, 657
Henson, W. E.: 310
Hess, W. J.: 311
High, G.: 312, 313, 314
Hill, K. T.: 659
Hill, P. S. M.: 315, 316
Hindman, F. G.: 317
Hirst, M. E.: 318
Hoffman, M. G.: 319
Hoisington, P. B.: 320
Holland, R.: 567
Holt, J. R.: 321, 322
Holzinger, J. M.: 323
Honess, C. W.: 324
Hopkins, M.: 325, 326, 327, 328, 329, 330
Hubbell, W. H.: 55
Huddleston, D. J.: 148
Huffine, W. H.: 331
Hulnik, F. W.: 332
Hunt, C. B.: 333
Hutcheson, H. L.: 334, 335
Hutchinson, G. P.: 336, 337
Irving, W.: 338, 339
James, E.: 340
Jameson, W. C.: 42
Jarman, R.: 341, 342
Jeffs, R. E.: 343
Jennison, M. C.: 555, 556
Jewell, J.: 247
Jex, G. W.: 222
Johnson, F. L.: 256, 344, 345, 346, 347, 348, 349, 350, 351, 352, 483, 630, 654, 655, 675
Johnson, J. C.: 353
Jones, R. E.: 354, 355, 356, 357
Jones, R. W.: 358
Jones, W.: 462
Kapustka, L. A.: 359, 360, 361, 362, 363, 364
Keck, D. W.: 365
Kelting, R. W.: 91, 366, 367, 368, 369, 370, 371, 372, 560, 610, 614
Kennedy, M. T.: 373, 374
Kennedy, R. K.: 375, 634, 635
Kessler, E.: 376, 377
Kingery, C. E.: 378
Kingsbury, P. J.: 379
Kirkham, M. B.: 377
Klinglesmith, R. E.: 380
Klopatek, J. M.: 381, 382, 383, 384
Koch, A. R.: 385
Koford, C. B.: 386
Kohlbacher, G. C.: 387
Kuchler, A. W.: 388, 389
Lahman, M. S.: 390
Landolt, J. C.: 391
Larkin, E. B.: 392
Larson, F.: 393
Lauenroth, W. K.: 687
Lawson, C. A.: 264, 265, 394, 395, 396
Leake, D. V.: 397
Leuck, E. E.: 398
Levetin, E. H.: 87, 399, 400, 401, 402
Lewis, J. C.: 487
Lewis, W. H.: 403
Lipscomb, B.: 404
Little, E. L., Jr.: 343, 405, 406, 407, 408, 409, 410, 411, 412, 413, 414,

415, 416
Lodhi, M. A. K.: 417, 418
Lofton, S. M.: 419
Long, A. W.: 420
Lottinville, S.: 511
Love, D.: 421
Love, H. S.: 422
Loveland, R. W.: 423
Lovell, J. F.: 424
Lowe, R. L.: 425
Lowry, D. P.: 585
Luckhardt, R. L.: 426
Lynch, R. A.: 427, 428, 570, 571
Mace, P.: 156
Magrath, L. K.: 429, 430, 431, 432, 433, 434, 463, 834
Mahler, W. F.: 435, 436
Mahmoud, I.: 247
Mallik, M. A. B.: 437, 438, 612
Maloney, M. M.: 439, 440
Mapp, L. R.: 441
Marcy, R. B.: 442, 443
Martin, R. E.: 444
Massey, J. R.: 264, 445
Masters, R. E.: 72
May, E. B.: 446
May, S. W.: 447, 637
McCollom, G. E.: 448
McCollum, T.: 449
McCoy, D.: 450, 451, 452, 453, 454, 455, 456
McCullough, C. Y.: 457
McCullough, R. B.: 458
McDermott, J. F.: 339
McDonald, C. B.: 459, 460
McElroy, C. A.: 461, 462
McGregor, R. L.: 463, 464, 465
McIlvane, E. H.: 466, 467, 468, 469, 470
McLeod, K. W.: 471, 472
McMillan, C.: 473
McMillen, W. N.: 474
McMurphy, W. E.: 271
McMurtry, W. M.: 475
McPherson, J. K.: 277, 284, 472, 476, 477, 566
Means, F. H.: 478
Mericle, L. W.: 479
Middleton, J. W.: 480
Milby, T. H.: 481, 482, 483, 484
Milton, W. H.: 485
Moleski, F. L.: 364, 486
Morrison, J. A.: 487
Morrison, T. H.: 360
Mueller, D. M. J.: 488
Mullen, W. C.: 489
Muller, C. H.: 490
Murray, C. L.: 491
Murray, D. S.: 71
Myers, W. S.: 492, 709
Navarro, J. A.: 493
Nease, F. R.: 494
Neill, R. L.: 495, 496
Nelson, C. D.: 497
Nelson, E. N.: 155, 498, 499
Nelson, G. L.: 500
Nelson, P. M.: 501
Nelson, R. L.: 502
Netherland, L.: 503
Neville, O.: 504
Nickell, G. L.: 505
Nighswonger, P. F.: 506, 507
Nixon, W. M.: 508
Nuttall, T.: 509, 510, 511
Olive, E. W.: 512
Olmstead, C. E.: 415, 416
Olmstead, C. E., Jr.: 513, 514
Olness, A.: 769
Omohundro, T. T.: 515
Ophel, I. L.: 516
Ortenberger, A. I.: 517, 518, 519
Orth, M. L. P.: 520
Osborn, B.: 22, 521, 522, 523
Osterhaus, C. A.: 524
Overmire, T. G.: 525
Paka, S.: 526
Palmer, E. J.: 527, 528, 529, 530
Pancholy, S. K.: 531, 532, 615, 616, 617
Parenti, R. L.: 533, 534, 613
Park, W. J.: 535
Parker, G. R.: 536
Parker, H. L.: 537
Parks, J. M.: 538, 539
Parton, W. J.: 540, 636, 637
Penfound, W. T.: 119, 145, 371, 372, 446, 484, 541, 542, 543, 544, 545, 546, 547, 548, 549, 550, 551, 552, 553, 554, 555, 556, 557, 558, 559, 560, 561, 610, 611, 618, 619, 620, 621, 673, 755, 808, 843
Pennington, R. E.: 658
Perino, C. H.: 561
Perino, J. V.: 562, 563
Perkins, G.: 564
Perkins, W. E.: 10

Petranka, J. W.: 565, 566, 567
Pfiester, L. A.: 148, 303, 322, 374, 428, 568, 569, 570, 571, 572, 857
Phillipi, M. A.: 573, 574, 780
Phillips, P. G.: 575, 576
Pireh, W.: 577, 578
Pollard, A. J.: 94
Polley, H. W.: 579, 580, 581
Porter, C. L.: 582, 583
Posey, C. E.: 584
Pound, R.: 585
Powell, J.: 277, 585
Preston, J. L.: 587
Preston, J. R.: 444
Prier, C. W.: 588
Proctor, V. W.: 275
Rasmussen, J. A.: 589, 590
Ray, H. E.: 591
Ray, R. J.: 592, 593
Reeb, J. E.: 594
Reed, L. W.: 298
Reeds, C. A.: 595
Reisen, W. K.: 596
Rhodes, L. K.: 597
Rice, E. L.: 3, 18, 58, 120, 140, 154, 212, 288, 361, 362, 363, 418, 438, 496, 514, 531, 532, 534, 557, 558, 559, 590, 598, 599, 600, 601, 602, 603, 604, 605, 606, 607, 608, 609, 610, 611, 612, 613, 614, 615, 616, 617, 618, 619, 620, 621, 632, 633, 719, 771, 850
Rice, M. O.: 622
Rinehart, F.: 462
Risk, P. L. W.: 623
Risser, P. G.: 6, 7, 140, 147, 253, 256, 257, 348, 349, 350, 351, 352, 382, 383, 384, 540, 563, 624, 625, 626, 627, 628, 629, 630, 631, 632, 633, 634, 635, 636, 637, 675, 678
Robertson, J. L.: 222
Rochelle, B. P.: 638
Rogers, C. M.: 639, 640
Rollins, R. C.: 641
Roozitalab, M. H.: 273
Rorbaugh, L. M.: 611
Rose, L. E.: 297
Ross, R. G.: 642
Roux, E. R.: 643
Rowland, N. W.: 832
Runyon, H. E.: 644, 650
Russell, C. S.: 230
Rutledge, J. M.: 645
Rydberg, P. A.: 646
Santlemann, P. W.: 173, 199
Savage, D. A.: 469, 470, 647, 648, 649, 650
Schaffner, J. F.: 651
Schemnitz, S. D.: 652, 653, 707
Schnell, G. D.: 654, 655, 656, 657
Segelquist, C. A.: 658
Sellards, E. H.: 659
Semtner, M. A.: 660
Semtner, P. J.: 661, 662, 663
Senter, W. R.: 664
Seyfer, J. R.: 665
Shannon, C. W.: 166, 666, 712
Shantz, H. L.: 667, 668, 669
Sharp, A. J.: 670, 671, 672
Shed, J. S.: 555, 556, 673
Sheedy, J. D.: 674, 675
Shelford, V. E.: 676
Sherwood, R. T. B.: 677, 678, 679
Sheviak, C. J.: 680, 681
Shirley, J. C.: 682, 683, 684
Siegler, D. S.: 685
Simison, B. D.: 846
Sims, P. L.: 686, 687, 688, 689, 690, 691
Singh, G.: 692
Singh, J. S.: 61, 687, 689, 690, 691
Skroch, J. A.: 693
Slosser, J. W.: 209
Smith, C. C.: 45, 694, 695, 696, 697
Smith, E. B.: 698
Smith, J. L.: 699
Smith, M.: 152
Snook, E. C.: 700
Sorenson, W.: 288
Spaulding, G. F.: 701
Spess, R. D.: 702
Springer, T. L.: 703, 704
Stanford, J. W.: 705
Stanlake, G. J.: 706
Stebler, A. M.: 707
Steffey, O. D.: 708
Stemen, R. R.: 709
Stemen, T. R.: 710
Stevens, G. W.: 166, 711, 712
Still, K. R.: 713
Stratton, R.: 714
Strine, J. H.: 715
Sullivan, J. R.: 449, 716
Svejcar, T.: 717
Swartz, P.: 158, 718
Tarr, J.: 120, 719

Tate, R. C.: 720
Taylor, C. E. S.: 721, 722, 723, 724, 725, 726, 727, 737, 738, 739, 740, 741, 742, 743, 744, 745, 746, 747, 748, 749, 750, 751, 752, 753, 754, 756
Taylor, R. J.: 312, 313, 314, 434, 723, 724, 725, 726, 728, 729, 730, 731, 732, 733, 734, 735, 736, 737, 738, 739, 740, 741, 742, 743, 744, 745, 746, 747, 748, 755, 756
Terry, S.: 572
Tharp, B. C.: 659
Thieret, J. W.: 757, 758, 759
Thomas, W. W.: 760
Thompson, G. A.: 403
Thompson, G. L.: 476, 761
Thompson, S.: 148
Thompson, W. H.: 762
Thorp, R. W.: 564
Todd, R. L.: 763
Torrey, J.: 764
Troeger, W. W.: 765, 766, 767, 768, 769
Turner, J. A.: 532, 770, 771
Tyler, J. D.: 772, 773
Tyrl, R. J.: 71, 94, 95, 96, 97, 98, 99, 100, 101, 102, 103, 104, 105, 106, 107, 108, 109, 110, 111, 202, 218, 219, 220, 221, 253, 477, 498, 573, 578, 631, 700, 704, 774, 775, 776, 777, 778, 779, 780, 781, 859
U. S. Army, Corps of Engineers: 782, 783, 784
U. S. Geological Survey: 785
U. S. War Dept.: 786, 787, 788, 789
Umber, R. L.: 790
Ungar, I. A.: 791, 792
U. S. Department of Agriculture: 793, 794, 795
University of Tulsa, Faculty of Natural Sciences: 796
Uno, G. E.: 142, 143, 797
Uttal, L. J.: 798
Van Vleet, A. H.: 799
Vaughn, J. M.: 800
Vowell, J. L.: 801
Wallace, L. L.: 11, 580
Wallis, C. S.: 802, 803, 804, 805, 825, 826
Ward, C. G.: 806
Ware, G. H.: 807, 808
Warren, M.: 643
Waterfall, U. T.: 330, 805, 809, 810, 811, 812, 813, 814, 815, 816, 817, 818, 819, 820, 821, 822, 823, 824, 825, 826
Watkins, W. E.: 827
Watson, L. J.: 828
Watson, S. H.: 829
Weaver, J. E.: 20, 830, 831, 832
Webb, R.: 158, 718
Webb, W. L.: 833
Weedon, R. R.: 463, 834
Weese, A. O.: 835, 836
Welborne, F. F.: 837
Welch, P.: 838
Wheelock, T.: 839
Whipple, A. W.: 789
White, P. J.: 840
Whitenton, R. O.: 841
Wiedeman, V. E.: 842, 843
Wiens, J. A.: 637
Wihlm, J.: 665
Willemsen, R. W.: 844
Williams, J. E.: 845
Williams, Q.: 474
Williams, S. T.: 846
Willson, D. L.: 847, 848
Wilson, R. E.: 849, 850
Wilson, W. D.: 851
Wingo, R. D.: 360
Wolfe, J. M.: 852
Woods, E. D.: 853
Wright, K. L. T.: 854
Wright, T. L.: 570, 571, 855, 856, 857
Zanoni, T. A.: 631, 858, 859, 860
Zon, R.: 669
Zwick, F. B.: 861, 862

COUNTY INDEX

Adair: 52
Alfalfa: 33, 113
Atoka: 51
Beaver: 354, 356, 357
Beckham: none
Blaine: 569
Bryan: 51, 723, 724, 728, 766, 767, 769
Caddo: 183, 184, 410, 411, 482, 837
Canadian: 67, 139, 183, 184, 245, 246, 411, 482, 600, 601, 717, 722, 837
Carter: 168, 169, 244, 328, 595
Cherokee: 52
Choctaw: 51, 783
Cimarron: 381, 639, 640, 720
Cleveland: 16, 19, 38, 45, 47, 65, 67, 308, 309, 345, 350, 351, 352, 361, 362, 366, 381, 392, 427, 428, 437, 438, 486, 507, 541, 544, 546, 582, 583, 588, 589, 590, 619, 620, 645, 763, 807, 849, 855, 857
Coal: none
Comanche: 84, 85, 90, 130, 142, 144, 162, 163, 183, 184, 186, 216, 217, 225, 285, 286, 319, 378, 411, 492, 500, 554, 580, 589, 590
Cotton: none
Craig: none
Creek: 293
Custer: 479
Delaware: 52, 390
Dewey: none
Ellis: none
Garfield: 475, 682
Garvin: none
Grady: 445
Grant: none
Greer: 36, 92
Harmon: 36, 444, 445
Harper: none
Haskell: 51
Hughes: 664
Jackson: 36
Jefferson: none
Johnston: 295
Kay: 116, 475, 485, 776, 784
Kingfisher: none
Kiowa: 34, 445
Latimer: 51
Le Flore: 279, 347
Lincoln: 67, 475
Logan: 67, 358, 440
Love: none
Major: none
Marshall: 9, 10, 38, 69, 346, 547, 553, 564, 599, 721, 855
Mayes: 52, 570, 571
McClain: 13, 14, 121, 212, 360, 363, 377, 445, 506, 537, 551, 557, 562, 563, 582, 588, 598, 611, 613
McCurtain: 51, 129, 151, 154, 381, 383, 412, 413, 602
McIntosh: 52
Murray: 42, 168, 169, 170, 171, 244, 328, 335, 595, 596
Muskogee: 52, 406, 408, 409, 601
Noble: none
Nowata: none
Okfuskee: none
Oklahoma: 810
Okmulgee: 439, 827
Osage: 61, 146, 147, 185, 198, 248, 250, 251, 304, 305, 306, 354, 356, 447, 504, 540, 625, 626, 627, 628, 629, 634, 635, 636, 674, 675, 687, 689, 690, 691, 776, 784
Ottawa: 52, 254, 255, 257, 429
Pawnee: 295
Payne: 48, 67, 149, 153, 175, 177, 181, 199, 228, 258, 271, 295, 296, 331, 336, 448, 503, 535, 536, 586, 660, 688, 692, 693
Pittsburg: none
Pontotoc: 455
Pottawatomie: 17, 43, 44, 67, 562, 563
Pushmataha: 51, 730
Roger Mills: none
Rogers: none
Seminole: none
Sequoyah: 52
Stephens: none
Texas: 38, 782
Tillman: none
Tulsa: 88, 89, 93, 133, 315, 316, 401, 402, 420, 504
Wagoner: none
Washington: 459, 460
Washita: none
Woods: 445, 679
Woodward: 467

SECTION-OF-STATE INDEX

Panhandle: 31, 115, 135, 193, 237, 317, 354, 356, 357, 474, 517, 639, 640, 653, 707, 720, 782

Northwest: 33, 113, 128, 188, 189, 193, 209, 242, 245, 246, 289, 292, 311, 349, 381, 445, 457, 466, 467, 468, 469, 470, 474, 479, 482, 517, 519, 522, 569, 600, 601, 652, 677, 679, 717, 791, 792, 842, 843

North central: 5, 6, 7, 12, 23, 27, 28, 40, 48, 50, 54, 61, 63, 67, 81, 116, 139, 140, 146, 147, 149, 153, 157, 161, 164, 174, 175, 177, 178, 181, 183, 184, 185, 198, 199, 206, 208, 210, 224, 228, 229, 233, 248, 250, 251, 252, 258, 263, 271, 277, 278, 284, 295, 296, 297, 304, 305, 306, 331, 334, 336, 337, 349, 354, 355, 356, 358, 365, 422, 440, 447, 448, 475, 476, 485, 488, 503, 504, 515, 524, 525, 535, 536, 540, 565, 566, 586, 603, 605, 607, 625, 626, 627, 628, 629, 634, 635, 660, 665, 674, 675, 682, 686, 687, 688, 689, 690, 691, 692, 693, 694, 696, 697, 710, 722, 761, 784, 810, 819, 837

Northeast: 41, 52, 53, 54, 72, 83, 87, 88, 89, 91, 93, 117, 133, 138, 224, 231, 233, 254, 255, 256, 257, 266, 293, 299, 315, 316, 348, 349, 390, 394, 401, 402, 406, 408, 409, 420, 423, 426, 429, 439, 458, 459, 460, 491, 504, 509, 511, 518, 528, 532, 570, 571, 592, 593, 601, 636, 654, 655, 661, 662, 663, 670, 702, 796, 802, 803, 804, 805, 825, 826, 827, 853

Southeast: 1, 24, 39, 51, 53, 72, 129, 137, 151, 154, 159, 172, 173, 187, 207, 211, 222, 233, 239, 253, 256, 279, 299, 312, 313, 324, 347, 348, 349, 381, 383, 391, 394, 403, 412, 413, 415, 416, 423, 458, 478, 489, 502, 509, 511, 518, 526, 527, 528, 594, 602, 615, 616, 617, 631, 638, 654, 658, 670, 672, 698, 702, 730, 731, 732, 733, 734, 735, 736, 742, 743, 744, 745, 746, 747, 748, 749, 750, 751, 752, 753, 754, 760, 783, 801, 851, 853

South central: 9, 10, 11, 13, 14, 16, 17, 18, 19, 21, 38, 42, 43, 44, 45, 47, 65, 69, 76, 78, 82, 119, 121, 122, 123, 124, 125, 141, 145, 152, 168, 169, 170, 171, 212, 213, 233, 239, 244, 247, 263, 267, 268, 269, 275, 278, 281, 282, 288, 295, 308, 309, 311, 320, 325, 328, 335, 341, 345, 346, 349, 350, 351, 352, 353, 360, 361, 362, 363, 364, 366, 367, 368, 369, 370, 371, 377, 380, 381, 391, 392, 397, 417, 418, 422, 424, 428, 437, 438, 441, 445, 446, 455, 456, 461, 462, 483, 486, 494, 495, 496, 505, 506, 507, 514, 524, 533, 537, 538, 539, 541, 544, 546, 547, 549, 550, 551, 552, 553, 555, 556, 557, 558, 560, 562, 563, 564, 567, 573, 575, 576, 579, 581, 582, 583, 588, 589, 590, 595, 596, 598, 599, 603, 607, 610, 611, 613, 615, 616, 617, 618, 619, 620, 643, 645, 664, 673, 694, 696, 697, 710, 718, 721, 723, 724, 728, 729, 755, 758, 761, 763, 766, 767, 769, 770, 807, 808, 838, 847, 848, 849, 850, 855, 857

Southwest: 22, 34, 35, 36, 84, 85, 90, 92, 130, 142, 144, 162, 163, 165, 183, 184, 186, 191, 192, 216, 217, 225, 239, 243, 252, 285, 286, 292, 314, 319, 341, 349, 378, 385, 391, 405, 410, 411, 444, 445, 466, 468, 474, 482, 487, 492, 500, 517, 523, 554, 579, 580, 581, 589, 590, 652, 659, 725, 772, 773, 797, 836, 837, 842, 843

All: 2, 3, 9, 10, 15, 16, 17, 18, 29, 38, 48, 55, 57, 58, 59, 60, 69, 75, 79, 86, 87, 88, 89, 114, 118, 131, 134, 166, 180, 194, 195, 220, 226, 230, 272, 301, 307, 333, 342, 343, 346, 352, 359, 362, 372, 379, 382, 388, 399, 400, 401, 402, 407, 414, 417, 418, 430, 450, 451, 454, 464, 465, 480, 481, 484, 486, 495, 496, 499, 513, 514, 521, 533, 534, 538, 547, 548, 553, 564, 585, 589, 590, 597, 603, 604, 607, 608, 609, 612, 615, 616, 617, 621, 622, 632, 633, 666, 671, 683, 684, 699, 709, 711, 712, 770, 771, 785, 793, 794, 795, 799, 824, 833, 835, 840, 841, 845, 849, 850, 852, 855, 859, 860

DUCK AND FLETCHER TYPE INDEX

Cypress bottoms forest: 413
Sand-sage grassland: 677, 679, 706
Shinnery oak-grassland: 444, 490, 706, 842, 843
Shortgrass high plains: 127, 464, 465, 679
Stabilized dune: 677, 679
Tallgrass prairie: 48, 464, 465
Mesquite grassland: 234, 444, 706
Oak-hickory forest: 347, 600, 601, 706
Post oak-blackjack: 344, 345, 349, 350, 351, 352, 619, 621
Piñon-juniper woodland: 639, 640
Mixedgrass eroded plains: none
Oak-pine forest: 347
Bottomland forest: 140, 346, 567, 605, 620

OTHER VEGETATION TYPE INDEX

Aquatic: 47, 113, 145, 148, 149, 150, 154, 155, 156, 178, 181, 275, 303, 322, 341, 358, 374, 379, 397, 427, 428, 499, 516, 542, 543, 568, 762, 765, 766, 767, 768, 769, 791, 856, 857, 860
Grassland: 2, 3, 8, 9, 11, 12, 13, 14, 22, 25, 27, 29, 31, 45, 48, 54, 56, 61, 77, 78, 79, 81, 82, 83, 90, 91, 120, 121, 122, 123, 124, 125, 130, 138, 139, 142, 144, 146, 147, 164, 170, 171, 176, 186, 197, 198, 204, 210, 219, 220, 224, 231, 247, 248, 249, 250, 251, 271, 288, 291, 298, 301, 304, 305, 306, 317, 331, 334, 335, 336, 338, 362, 363, 367, 369, 376, 377, 378, 384, 386, 389, 419, 423, 447, 461, 469, 470, 474, 475, 485, 488, 491, 494, 496, 503, 505, 508, 514, 515, 523, 525, 535, 537, 540, 551, 556, 558, 560, 562, 563, 576, 580, 581, 583, 585, 586, 587, 592, 593, 599, 607, 611, 612, 613, 615, 616, 617, 618, 624, 625, 626, 627, 628, 629, 634, 635, 636, 637, 643, 646, 647, 648, 649, 650, 653, 660, 664, 668, 674, 675, 686, 688, 692, 693, 696, 697, 713, 717, 719, 720, 761, 806, 829, 830, 836, 838, 847, 848, 860
Forest: 1, 5, 6, 7, 21, 23, 30, 40, 49, 56, 72, 79, 84, 85, 90, 93, 119, 132, 134, 159, 171, 186, 187, 190, 199, 200, 207, 211, 213, 223, 229, 235, 238, 267, 274, 278, 281, 282, 283, 284, 292, 293, 294, 297, 299, 302, 309, 315, 316, 318, 320, 324, 332, 335, 345, 347, 350, 410, 411, 413, 414, 415, 416, 420, 422, 424, 426, 458, 482, 489, 502, 524, 526, 527, 541, 567, 594, 600, 601, 602, 615, 616, 617, 619, 620, 621, 646, 656, 657, 708, 715, 729, 760, 801, 837, 839, 845, 847, 851, 860
Pine: 706, 728
Prairie: 22, 25, 27, 28, 29, 45, 56, 77, 78, 79, 82, 124, 126, 127, 130, 142, 144, 157, 162, 163, 170, 176, 186, 198, 200, 204, 218, 219, 220, 248, 250, 251, 274, 288, 291, 298, 301, 302, 304, 305, 306, 320, 334, 335, 337, 352, 362, 370, 384, 462, 470, 473, 474, 494, 505, 535, 550, 552, 554, 560, 575, 576, 579, 580, 581, 583, 586, 587, 598, 608, 615, 616, 617, 624, 625, 626, 627, 628, 629, 630, 634, 635, 648, 649, 650, 651, 660, 668, 674, 686, 687, 689, 690, 691, 692, 693, 706, 717, 718, 755, 761, 797, 806, 831, 832, 836, 847, 848

SUBJECT-IN-CONTEXT INDEX

ABANDONED Revegetation of abandoned fields in Kansas and Oklahoma. Amer. J. 67
ABANDONED A comparison of the soil fungi of a tall-grass prairie and of a 212
ABANDONED A comparison of the soil fungi of a virgin prairie and an abandon 461
ABANDONED Data and observations concerning regrassing abandoned cultivated 475
ABANDONED Vegetation of an abandoned prairie dog town in tall-grass prairie 523
ABANDONED Soil factors preventing revegetation of a denuded area near an ab 532
ABANDONED A comparison of the insects of a virgin prairie, a grazed pasture 537
ABANDONED Inhibition of nodulation of inoculated legumes by pioneer plant s 607
ABANDONED Seed dispersal and mineral nutrition in succession in abandoned f 611
ABANDONED Plant succession on abandoned fields in central Oklahoma and in t 643
ABANDONED Natural revegetation of abandoned farmland in the central and sou 650
ABANDONED Flora and fauna of Oklahoma abandoned mine lands. Okla. Biol. Sur 654
ABANDONED Biotic and physiographic succession on abandoned eroded farmland. 695
ABANDONED Biotic and physiographic succession on abandoned eroded farmland. 697

ABIOTIC The structure and function of ten western North American grasslands 687

ACER Relationships of two isolated groups of sugar maples (Acer saccharum M 184
ACER Relationships of sugar maples (Acer saccharum and A. grandidentatum) i 252
ACER Acer grandidentatum in Oklahoma. Rhodora 46: 445-450. {SW Canadian Cad 411
ACER Germination studies of "Caddo" sugar maple (Acer saccharum) seeds. See 482
ACER The microclimate of sugar maple stands in Oklahoma. Ecology 43: 19-25. 601

ACETYLENE Acetylene reduction (nitrogen fixation) and old-field succession 359
ACETYLENE Acetylene reduction (N-fixation) of glucose amended soils from ce 361
ACETYLENE Acetylene reduction (N-fixation) in soil and old-field succession 362

ACHILLEA Cytogeography and reproductive biology of Achillea millefolium in 577
ACHILLEA Cytogeography of Achillea millefolium in Oklahoma and adjacent sta 578
ACHILLEA Origin and distribution of polyploid Achillea (Compositae) in west 774
ACHILLEA The genus Achillea. In: A synonimized checklist of the vascular fl 777
ACHILLEA Nomenclature of the genus Achillea. In: National list of scientifi 778

ACID Inhibition of symbiotic nitrogen-fixation by gallic and gallotannic ac 57
ACID Inhibition of symbiotic nitrogen fixation by gallic and tannic acid, a 58
ACID Loss of ferulic acid from decomposing leaves of Celtis laevigata and s 770
ACID Microbial decomposition of ferulic acid in soil. J. Chem. Ecol. 1: 41- 771

AEROALLERGENS Aeroallergens in Tulsa, Oklahoma. Sixth Intl. Palynological C 401

AFFINITIES The floristic affinities of the Arbuckle Mountains in Oklahoma. 328

AFTER-RIPENING After-ripening and germination of seed of Juniperus virginia 160

AGRICULTURE Natural vegetation. Atlas of American Agriculture, Sec. E: 1-29 669
AGRICULTURE Soils and men. Yearbook of agriculture. U. S. Govt. Printing Of 793
AGRICULTURE Climate and man. Yearbook of agriculture. U. S. Govt. Printing 794
AGRICULTURE Trees. Yearbook of agriculture. U. S. Govt. Printing Office, Wa 795

ALETRIS A numerical analysis of the morphological variation in the genus Al 623

ALFISOL The Gulf Coastal Plain Ultisol and Alfisol ecosystem: surface geolo 502

ALGA Water quality and phytoplankton communities in Lake Thunderbird, Cleve 47
ALGA Algae as pioneers in plant succession and their importance in erosion 66
ALGA The effect of salinity on the distribution of algae at the Salt Plains 113
ALGA A third species of Enteromorpha (a marine Chlorophycean) for Oklahoma. 148
ALGA Some algae of Lake Carl Blackwell, Oklahoma, exclusive of diatoms. M. 149

ALGA Spatial and temporal variation in productivity, species diversity, and 150
ALGA Vegetative habit and reproduction of Desmidium grevillii (Kütz) de Bar 154
ALGA A quantitative comparison of the soil algae of central Oklahoma prairi 213
ALGA Electrophoretic analysis of soluble enzymes in five freshwater dinofla 303
ALGA Studies in the dinoflagellate genera, Peridinium and Peridiniopsis. M. 321
ALGA A survey of auxotrophy in five freshwater dinoflagellates (Pyrrhophyta 322
ALGA Analysis of the diatom (Bacillariophyta) communities of the North Cana 373
ALGA Additions to the diatom (Bacillariophyceae) flora of Oklahoma. Proc. O 374
ALGA Plankton production and species distribution in the limnological provi 379
ALGA Diatoms, including salt-water taxa, from southwestern Oklahoma. Proc. 385
ALGA The algae of Crystal Lake, Cleveland County, Oklahoma. Amer. Midl. Nat 397
ALGA The algae of Crystal Lake, 1938-1978. M. S. Thesis, Univ. Okla., Norma 427
ALGA Crystal Lake revisited. Proc. Okla. Acad. Sci. 61: 49-57. {SC Clevelan 428
ALGA Seasonal algal flora on the campus of Catholic College, Guthrie, Oklah 440
ALGA A limnological study of two artificial lakes in Oklahoma. M. S. Thesis 516
ALGA Effects of certain plants of old-field succession on the growth of blu 538
ALGA Coronastrum: a rare find. Trans. Amer. Micros. Soc. 96: 163. {aquatic 568
ALGA Enteromorpha, a marine alga in Oklahoma. Proc. Okla. Acad. Sci. 56: 66 569
ALGA Species composition and diversity of phytoplankton in the Grand River 570
ALGA Species composition and diversity of periphyton in the Grand River Dam 571
ALGA Additions to the algae of Oklahoma. Southw. Natur. 23: 85-94. {} 572
ALGA The ecology of Honey Creek: Temporal patterns of the travertine periph 596
ALGA Variation with stream order in species composition, diversity, biomass 665
ALGA Lemnaea fuscina Borg (Rhodophyta) in Oklahoma. Proc. Okla. Acad. Sci. 730
ALGA A limnological study of two small artificial lakes. M. S. Thesis, Univ 762
ALGA Epiphytic diatoms as indicators of water quality in farm and experimen 765
ALGA Survey of algae in farm ponds in Bryan County, Oklahoma. Southw. Natur 766
ALGA Epiphytic diatoms in farm ponds and experimental ponds in Bryan County 767
ALGA Epiphytic diatoms as indicators of water quality in farm and experimen 768
ALGA Plant development and sequence in artificial ponds with and without ph 769
ALGA A survey of the soil algae of two vegetation types in central Oklahoma 847
ALGA An exploratory study on soil algae. Ecology 38: 309-313. {SC grassland 848
ALGA Allelopathic interactions between algae. M. S. Thesis, Univ. Okla., No 852
ALGA Gloeodendron: A rare alga in Oklahoma. Proc. Okla. Acad. Sci. 57: 169- 855
ALGA A comparison of algal succession between farm ponds with sandstone and 856
ALGA A comparison of algal composition and seasonal succession between pond 857

ALLELOPATHIC Allelopathic effects of Platanus occidentalis L. M. S. Thesis, 16
ALLELOPATHIC Allelopathic effects of Platanus occidentalis L. in relation t 17
ALLELOPATHIC Allelopathic effects of Platanus occidentalis. Bull. Torrey Bo 18
ALLELOPATHIC Allelopathic effects of Polygonum aviculare L. Ph. D. Thesis, 19
ALLELOPATHIC Allelopathic effects of Celtis laevigata Willd. as related to 417
ALLELOPATHIC Allelopathic effects of Celtis laevigata. Bull. Torrey Bot. Cl 418
ALLELOPATHIC Competitive and allelopathic suppression of understory by Okla 476
ALLELOPATHIC Allelopathic effects of Sporobolus pyramidatus on vegetational 589
ALLELOPATHIC Allelopathic effects of Sporobolus pyramidatus on vegetational 590
ALLELOPATHIC Some roles of allelopathic compounds in plant communities. Bio 609
ALLELOPATHIC Allelopathic inhibition of weed invasion by Andropogon scopari 713
ALLELOPATHIC Allelopathic interactions between algae. M. S. Thesis, Univ. O 852

ALLELOPATHY Plant inhibition by Johnson grass and its possible significance 2
ALLELOPATHY Plant inhibition by Johnson grass and its possible significance 3
ALLELOPATHY Ecological studies of plant inhibition by Solanum rostratum. Ph 15

ALLELOPATHY Allelopathic effects of Platanus occidentalis L. M. S. Thesis, 16
ALLELOPATHY Allelopathic effects of Platanus occidentalis L. in relation to 17
ALLELOPATHY Allelopathic effects of Platanus occidentalis. Bull. Torrey Bot 18
ALLELOPATHY Inhibition of symbiotic nitrogen-fixation by gallic and gallota 57
ALLELOPATHY Inhibition of symbiotic nitrogen fixation by gallic and tannic 58
ALLELOPATHY The possible ecological significance of inhibition of Euphorbia 75
ALLELOPATHY Acetylene reduction (nitrogen fixation) and old-field successio 359
ALLELOPATHY Acetylene reduction (N-fixation) in soil and old-field successi 362
ALLELOPATHY Allelopathic effects of Celtis laevigata Willd. as related to p 417
ALLELOPATHY Allelopathic effects of Celtis laevigata. Bull. Torrey Bot. Clu 418
ALLELOPATHY Condensed tannins in soil: inputs and effects on microbial popu 486
ALLELOPATHY Possible role of Ambrosia psilostachya on patterning and succes 495
ALLELOPATHY Possible role of Ambrosia psilostachya on pattern and successio 496
ALLELOPATHY Effects of bacterial inhibitors and certain related compounds o 513
ALLELOPATHY Relative effects of known plant inhibitors on species from firs 514
ALLELOPATHY Inhibitional effects of Digitaria sanguinalis and possible role 533
ALLELOPATHY Inhibitional effects of Digitaria sanguinalis and possible role 534
ALLELOPATHY Effects of certain plants of old-field succession on the growth 538
ALLELOPATHY Allelopathic effects of Sporobolus pyramidatus on vegetational 589
ALLELOPATHY Allelopathic effects of Sporobolus pyramidatus on vegetational 590
ALLELOPATHY Inhibition of nitrogen-fixing and nitrifying bacteria by seed p 603
ALLELOPATHY Inhibition of nitrogen-fixing and nitrifying bacteria by seed p 604
ALLELOPATHY Inhibition of nodulation of inoculated legumes by pioneer plant 607
ALLELOPATHY Allelopathy and grassland improvement. In: Estes, J. R., and R. 608
ALLELOPATHY Some roles of allelopathic compounds in plant communities. Bioc 609
ALLELOPATHY Causes of decreases in residual carbohydrase activity in soil d 612
ALLELOPATHY Inhibition of nitrification by climax ecosystems. Amer. J. Bot. 615
ALLELOPATHY Inhibition of nitrification by climax ecosystems. II. Additiona 616
ALLELOPATHY Inhibition of nitrification by climax ecosystems. III. Inhibito 617
ALLELOPATHY Differences in nitrate reductase activity between species of di 699
ALLELOPATHY Loss of ferulic acid from decomposing leaves of Celtis laevigat 770
ALLELOPATHY Microbial decomposition of ferulic acid in soil. J. Chem. Ecol. 771
ALLELOPATHY A study of allelopathy in Helianthus rigidus. M. S. Thesis, Okl 790
ALLELOPATHY Allelopathy as expressed by Helianthus annuus, and its role in 849
ALLELOPATHY Allelopathy as expressed by Helianthus annuus and its role in o 850
ALLELOPATHY Allelopathic interactions between algae. M. S. Thesis, Univ. Ok 852

ALLIUM Allium ampeloprasum (Liliaceae) and Trifolium vesiculosum (Leguminos 757

AMARYLLIDACEAE Biology and distribution of Hypoxis longii Fern. (Amaryllida 753

AMBROSIA Possible role of Ambrosia psilostachya on patterning and successio 495
AMBROSIA Possible role of Ambrosia psilostachya on pattern and succession i 496
AMBROSIA Mechanism of seed dormancy in Ambrosia artemisiifolia. Ph. D. Diss 844

AMORPHA Amorpha ouachitensis. Status report. Dept. of the Army, U. S. Army 734
AMORPHA Amorpha ouachitensis. Status report. Endangered Species Office, U. 743

ANDROPOGON Evidence for the absence of ecotypic development in Andropogon v 257
ANDROPOGON Carbon dioxide exchange characteristics of some prairie grass se 630
ANDROPOGON Allelopathic inhibition of weed invasion by Andropogon scoparius 713
ANDROPOGON Notes on the growth and fruiting of Andropogon scoparius, with s 806

ANGIOSPERMS A study of the angiosperms, exclusive of the Gramineae, Cyperac 153
ANGIOSPERMS A partial list of the angiosperms of Garfield County. Proc. Okl 682

ANNUAL Drought years in central Oklahoma from 1710 to 1959 calculated from 295
ANNUAL Biomass, annual net primary production and dynamics of six mineral e 345
ANNUAL Correlation analysis of rainfall and annual ring index of central Ok 350
ANNUAL Biomass, annual net primary production, and dynamics of six mineral 351
ANNUAL Germination of downy brome from southern Kansas, central Oklahoma, a 483
ANNUAL Plant population changes in a native prairie plot plowed annually ov 558

AQUATIC Water quality and phytoplankton communities in Lake Thunderbird, Cl 47
AQUATIC The effect of salinity on the distribution of algae at the Salt Pla 113
AQUATIC The relation of stock pond vegetation to geological formations in M 145
AQUATIC A third species of Enteromorpha (a marine Chlorophycean) for Oklaho 148
AQUATIC Some algae of Lake Carl Blackwell, Oklahoma, exclusive of diatoms. 149
AQUATIC Spatial and temporal variation in productivity, species diversity, 150
AQUATIC Vegetative habit and reproduction of Desmidium grevillii (Kütz) de 154
AQUATIC Myriophyllum in Oklahoma. Proc. Okla. Acad. Sci. 63: 103-104. {aqua 155
AQUATIC Eurasian water milfoil in Oklahoma. Proc. Okla. Acad. Sci. 58: 4-5. 156
AQUATIC Water fluctuations as a factor in the life of six of the higher pla 178
AQUATIC A preliminary study of the larger aquatic plants of Oklahoma with s 179
AQUATIC A preliminary study of the larger aquatic plants of Oklahoma with s 180
AQUATIC Water fluctuation as a factor in the life of the higher plants of a 181
AQUATIC A guide and key to the aquatic plants of the southeastern United St 222
AQUATIC A population study of Chara zeylanica in Texas, Oklahoma, and New M 275
AQUATIC Chemical composition of certain aquatic plants. Bot. Gaz. 96: 186-1 296
AQUATIC Electrophoretic analysis of soluble enzymes in five freshwater dino 303
AQUATIC A survey of auxotrophy in five freshwater dinoflagellates (Pyrrhoph 322
AQUATIC Eurasian watermilfoil - A new menace to Oklahoma waters. Proc. Okl 341
AQUATIC The development of aquatic ecoregions in Oklahoma. Ph. D. Thesis, U 342
AQUATIC Observations on the effects of erosion on the ecology and maintenan 358
AQUATIC Additions to the diatom (Bacillariophyceae) flora of Oklahoma. Proc 374
AQUATIC Plankton production and species distribution in the limnological pr 379
AQUATIC The algae of Crystal Lake, Cleveland County, Oklahoma. Amer. Midl. 397
AQUATIC The algae of Crystal Lake, 1938-1978. M. S. Thesis, Univ. Okla., No 427
AQUATIC Crystal Lake revisited. Proc. Okla. Acad. Sci. 61: 49-57. {SC Cleve 428
AQUATIC Aquatic plants of Oklahoma: I. Submersed, floating-leaved, and sele 499
AQUATIC A limnological study of two artificial lakes in Oklahoma. M. S. The 516
AQUATIC An outline for ecological life histories of herbaceous vascular hyd 542
AQUATIC Plant communities of Oklahoma lakes. Ecology 34: 561-583. {aquatic} 543
AQUATIC Primary production of vascular aquatic plants. Limnol. and Oceanog. 545
AQUATIC Coronastrum: a rare find. Trans. Amer. Micros. Soc. 96: 163. {aquat 568
AQUATIC Epiphytic diatoms as indicators of water quality in farm and experi 765
AQUATIC Survey of algae in farm ponds in Bryan County, Oklahoma. Southw. Na 766
AQUATIC Epiphytic diatoms in farm ponds and experimental ponds in Bryan Cou 767
AQUATIC Epiphytic diatoms as indicators of water quality in farm and experi 768
AQUATIC Plant development and sequence in artificial ponds with and without 769
AQUATIC Salinity tolerance of marsh plants in Kansas and Oklahoma. Ecology 791
AQUATIC A comparison of algal succession between farm ponds with sandstone 856
AQUATIC A comparison of algal composition and seasonal succession between p 857
AQUATIC Natural areas for Oklahoma: important areas of natural significance 860

ARALIACEAE Panax quinquefolium (Araliaceae) in southeastern Oklahoma. Rhodo 403

ARBORESCENT The arborescent flora of Oklahoma. M. S. Thesis, Univ. Wyo., La 134

ARBUCKLE A preliminary survey of the flora of the Arbuckle Mountains. M. S. 168
ARBUCKLE A preliminary survey of the flora of the Arbuckle Mountains. Texas 169

ARBUCKLE A preliminary list of the plants of the Arbuckle Mountains. B. A. 244
ARBUCKLE Variation and hybridization in Juniperus. Ann. Mo. Bot. Gard. 39: 281
ARBUCKLE The floristic affinities of the Arbuckle Mountains in Oklahoma. Am 328
ARBUCKLE Vegetation in relation to slope exposure and geology in the Arbuck 335
ARBUCKLE The Arbuckle Mountains, Oklahoma. Okla. Geol. Surv. Circ. No. 14. 595

ARCHAEOLOGICAL Identification and mapping of the extant flora at the Deer C 776

ARISTIDA Systematics of Aristida dichotoma, basiramea, and curtisii (Poacea 800

ARKANSAS Journal of travels into the Arkansas Territory during the year 181 509
ARKANSAS Collections towards a flora of the Territory of Arkansas. Trans. A 510
ARKANSAS A journal of travels into the Arkansas Territory during the year 1 511
ARKANSAS On Nuttall's trail through Arkansas. J. Arnold Arb. 8: 24-52. {SE 528

ARMS-LENGTH An evaluation of the arms-length rectangle method in forest sam 559

ARTEMISIA Biosystematics of Artemisia carruthii and Artemisia mexicana. M. 62
ARTEMISIA A biosystematic study of Artemisia caruthii. Ph. D. Dissertation, 591

ASCLEPIAS The geographic variation of Asclepias tuberosa interior in Oklaho 828

ASH An analysis of an elm-ash floodplain community near Norman, Oklahoma. P 541
ASH Composition of a green ash forest near Norman, Oklahoma. Southw. Natur. 620

ASPECTION Fluctuations in biotic communities. III. Aspection in a ravine se 124
ASPECTION Fluctuations in biotic communities. V. Aspection in mixed-grass p 125
ASPECTION Aspection in a central Oklahoma tension zone area. M. S. Thesis, 320

ASPECTS Seasonal aspects in certain prairie and savannah communities. M. S. 121
ASPECTS A preliminary report on the seasonal aspects of six habitats near N 308
ASPECTS Some aspects of structure and function in Oklahoma old-field succes 562
ASPECTS Some aspects of structure and function in Oklahoma old-field succes 563

ASSOCIATIONS The effect of host specificity on the interspecific associatio 7
ASSOCIATIONS Plant associations as site predictors in the pine-hardwood ten 187
ASSOCIATIONS Shortleaf pine site index – Soils and plant associations on t 851

ASTER A monographic study of the genus Aster in Oklahoma. M. S. Thesis, Uni 236

ASYMBIOTIC Symbiotic and asymbiotic N2-fixation in a tall grass prairie. So 363

ATLAS Atlas of the flora of the Great Plains. Iowa State Univ. Press, Ames. 464
ATLAS Natural vegetation. Atlas of American Agriculture, Sec. E: 1-29. {} 669
ATLAS Atlas of the woody plants of Oklahoma. Okla. Biol. Surv., Norman. {tr 845

AUTECOLOGICAL An autecological study of Salix nigra, especially factors tha 471

AUTUMN Oklahoma spring flowers in autumn. Proc. Okla. Acad. Sci. 17: 60-62. 325

AUXOTROPHY A survey of auxotrophy in five freshwater dinoflagellates (Pyrrh 322

BACCHARIS The genus Baccharis in the southwestern states of Oklahoma, Texas 435

BACILLARIOPHYCEAE Additions to the diatom (Bacillariophyceae) flora of Okla 374

BACILLARIOPHYTA Analysis of the diatom (Bacillariophyta) communities of the 373

BACTERIA Effects of bacterial inhibitors and certain related compounds on t 513
BACTERIA Inhibition of nitrogen-fixing and nitrifying bacteria by seed plan 603
BACTERIA Inhibition of nitrogen-fixing and nitrifying bacteria by seed plan 604

BALDCYPRESS Baldcypress (Taxodium distichum) in Oklahoma. Proc. Okla. Acad. 413

BARK Some factors influencing the frequency of bark lichens in north centra 5
BARK Some factors influencing the frequency of bark lichens in north centra 6
BARK The effect of host specificity on the interspecific associations of ba 7

BAUM Differential selection in juniper populations from the Baum Limestone 282
BAUM The grassland communities of the Baum Limestone in Johnston County, Ok 755

BEAVER'S BEND Vegetation of Beaver's Bend State Park. Okla. Geol. Surv. Gui 602

BEHAVIOR Habitat used by lesser prairie chickens for feeding related to sea 357
BEHAVIOR The ecology and behavior of the lone star tick (Acarina: Ixodidae) 661
BEHAVIOR The ecology and behavior of the lone star tick. (Acarina: Ixodidae 663

BIBLIOGRAPHY A bibliography of the Wichita Mountains of Oklahoma. Great Pla 286
BIBLIOGRAPHY Literature on the vegetation of Oklahoma. Proc. Okla. Acad. Sc 372
BIBLIOGRAPHY Literature on the vegetation of Oklahoma, 1964-1975. Proc. Okl 481
BIBLIOGRAPHY Additions to the literature on the vegetation of Oklahoma. Pro 484
BIBLIOGRAPHY Publications of the Oklahoma Biological Survey. Inf. Bull. No. 858

BIO-ECONOMIC Bio-economic analysis of even aged shortleaf pine stands in so 211

BIOGEOGRAPHIC A biogeographical boundary: The Tatschl Line. Trans. Kans. Ac 389
BIOGEOGRAPHIC Biogeographic regions of Texas and Oklahoma. Ecology 31: 427- 833

BIOGRAPHY Notes from the Bebb Herbarium of the University of Oklahoma II. R 329
BIOGRAPHY Umaldy T. Waterfall (1910-1971). Rhodora 73: 146-149. {biography 705

BIOLOGIC-ECONOMIC Eleven year biologic-economic evaluation of undesirable h 489

BIOMASS Nutrient and biomass allocation strategies for five grass species i 8
BIOMASS Nutrient and biomass allocation in five grass species in an Oklahom 11
BIOMASS Biomass, annual net primary production and dynamics of six mineral 345
BIOMASS Revegetation of unreclaimed coal strip-mines in Oklahoma. I. Vegeta 348
BIOMASS Biomass, annual net primary production, and dynamics of six mineral 351
BIOMASS Energy analyses of forestry practices in oak-pine forests and loblo 383
BIOMASS Effects of denudation on the productivity of grassland. Ecology 45: 552
BIOMASS Osage site, 1970 report: Primary production. USIBP Grassland Biome 626
BIOMASS Variation with stream order in species composition, diversity, biom 665
BIOMASS The structure and function of ten western North American grasslands 687
BIOMASS The structure and function of ten western North American grasslands 689
BIOMASS The structure and function of ten North American grasslands. III. N 690
BIOMASS The structure and function of ten western North American grasslands 691
BIOMASS The influence of burning on tillering and aboveground biomass parti 717

BIOME An assessment of weather patterns and ragweed pollen in Tulsa over a 89
BIOME Ecology of the grassland biome. M. S. Thesis, Univ. Okla., Norman. {p 126
BIOME The grassland biome. Ecol. Monogr. 10: 617-684. {prairie shortgrass} 127
BIOME Wind patterns and the seasonal occurrence of Juniperus pollen in Tuls 402
BIOME Preliminary producer data synthesis: 1970 comprehensive network sites 625
BIOME Osage site, 1970 report: Primary production. USIBP Grassland Biome Te 626
BIOME A preliminary compartment model of a tall grass prairie: Osage 1970. 628
BIOME Herbage dynamics of a tall grass prairie: Osage, 1971. USIBP Grasslan 634
BIOME Herbage dynamics of an Oklahoma tallgrass prairie: Osage 1972. USIBP 635

BIOSYSTEMATIC Biosystematics of Artemisia carruthii and Artemisia mexicana. 62
BIOSYSTEMATIC A biosystematic study of the relationship of Nama hispidum an 276
BIOSYSTEMATIC Biosystematic studies in the Echinocereus viridiflorus comple 398
BIOSYSTEMATIC A biosystematic study of Artemisia caruthii. Ph. D. Dissertat 591

BIOSYSTEMATIC Biosystematic study of the Spiranthes cernua complex. Bull. 4 680
BIOSYSTEMATIC A biosystematic study of the relationship of Nama hispidum an 779

BIOTA The effect of overgrazing and erosion upon the biota of the mixed gra 696

BIOTIC The biotic districts of Oklahoma. Amer. Midl. Natur. 20: 425-455. {c 55
BIOTIC Inhibition of symbiotic nitrogen-fixation by gallic and gallotannic 57
BIOTIC Inhibition of symbiotic nitrogen fixation by gallic and tannic acid, 58
BIOTIC Fluctuations in biotic communities. III. Aspection in a ravine sere 124
BIOTIC Fluctuations in biotic communities. V. Aspection in mixed-grass prai 125
BIOTIC The black-tailed prairie dog (Cynomys ludovicianus) as a member of a 176
BIOTIC Symbiotic and asymbiotic N2-fixation in a tall grass prairie. Soil B 363
BIOTIC Biotic type mapping of Oklahoma watersheds. Proc. Okla. Acad. Sci. 2 521
BIOTIC The structure and function of ten western North American grasslands. 687
BIOTIC Biotic and physiographic succession on abandoned eroded farmland. Ph 695
BIOTIC Biotic and physiographic succession on abandoned eroded farmland. Ec 697
BIOTIC Oklahoma climagraphs and biotic regions. Proc. Okla. Acad. Sci. 5: 9 835

BIRD Ecological relationships of the Bird Creek region, northeastern Oklaho 54
BIRD Ecology of the nesting birds of the McCurtain Game Preserve, Oklahoma. 129
BIRD Habitat preferences among representative wintering and breeding birds 353
BIRD A study of viable seeds in a selection of birds' nests. Ph. D. Thesis, 507

BISON Effects of disturbance on diversity in mixed-grass prairie. Vegetatio 144
BISON The role of the bison in maintaining the short grass plains. Ecology 393

BITTERWEEDS Toxic plants of Oklahoma: Bitterweeds and sneezeweeds. J. Okla. 98

BLACK-TAILED The black-tailed prairie dog (Cynomys ludovicianus) as a membe 176

BLACKJACK The short grass plains and post oak-blackjack woodland of Oklahom 56
BLACKJACK An evaluation of the quadrat method in the blackjack-post oak for 152
BLACKJACK A comparison of post oak-blackjack oak communities in two major s 199
BLACKJACK Biomass, annual net primary production and dynamics of six minera 345
BLACKJACK Correlation analysis of rainfall and annual ring index of central 350
BLACKJACK Biomass, annual net primary production, and dynamics of six miner 351
BLACKJACK An evaluation of the variable-radius and paired-tree methods in t 619

BLACKJACK-POST OAK An evaluation of the quadrat method in the blackjack-pos 152
BLACKJACK-POST OAK An evaluation of the variable-radius and paired-tree met 619

BLACKWELL Some algae of Lake Carl Blackwell, Oklahoma, exclusive of diatoms 149
BLACKWELL The drowning of trees in Lake Carl Blackwell. Proc. Okla. Acad. S 228

BLM A report on the threatened and endangered plant species in the BLM sout 631

BLUE-GREEN Effects of certain plants of old-field succession on the growth 538

BLUESTEM Vegetation-soil relationships in Flint Hills bluestem pastures. J. 27
BLUESTEM Analysis of morphological variation in Schizachyrium scoparium (li 77
BLUESTEM The influence of burning on tillering and aboveground biomass part 717

BOTANISTS Early botanists of Oklahoma. M. S. Thesis, Okla. A & M Coll., Sti 310

BOTANIZING Some results of a summer's botanizing in Oklahoma. Rhodora 51: 1 813
BOTANIZING Some results of a third summer's botanizing in Oklahoma. Rhodora 816

BOTTOMLAND Growth and distribution of Platanus occidentalis in Oklahoma as 30
BOTTOMLAND Bottomland hardwoods of eastern Oklahoma: A special study of the 72
BOTTOMLAND Ecological studies in an eastern Oklahoma floodplain. Proc. Okla 93

BOTTOMLAND Reproduction of woody plants in the absence of grazing. Proc. Ok 119
BOTTOMLAND Ordination and classification of mature bottomland forests in no 140
BOTTOMLAND Silting and forest succession on Deep Fork in southwestern Creek 229
BOTTOMLAND Effect of silting on tree development in the flood plain of the 293
BOTTOMLAND Effect of silt on natural vegetation and drainage in the flood p 297
BOTTOMLAND Ecological studies on the Canadian River floodplain in Cleveland 309
BOTTOMLAND Relation between soil fungi and seed plants in three successiona 438
BOTTOMLAND An autecological study of Salix nigra, especially factors that m 471
BOTTOMLAND Factors limiting the distribution of Salix nigra. Bull. Torrey B 472
BOTTOMLAND Edaphic and topographic effects on forest communities in Payne C 536
BOTTOMLAND An analysis of an elm-ash floodplain community near Norman, Okla 541
BOTTOMLAND The composition of a black willow forest in south central Oklaho 547
BOTTOMLAND A quantitative analysis of bottomland communities in south-centr 567
BOTTOMLAND A comparative study of upland and bottomland types of Panicum vi 582
BOTTOMLAND Bottomland forests of north-central Oklahoma. Ecology 46: 708-71 605
BOTTOMLAND Composition of a green ash forest near Norman, Oklahoma. Southw. 620
BOTTOMLAND Some variations of the fruit of Quercus macrocarpa. M. S. Thesis 708
BOTTOMLAND The vegetation of the lower levels of the South Canadian River i 808
BOTTOMLAND Journal of the expedition of the regiment of dragoons into India 839

BOUTELOUA Some morphological and physiological effects of density and densi 197
BOUTELOUA Competitive responses of Bouteloua curtipendula (Michx.) Torr. Am 624
BOUTELOUA Carbon dioxide exchange characteristics of some prairie grass see 630

BREEDING Habitat preferences among representative wintering and breeding bi 353

BROWSE The influence of varying light intensities on the vegetative growth 49
BROWSE Growth of ten species of preferred deer browse plants under various 132
BROWSE Vegetation response following winged elm (Ulmus alata Michx.) and oa 172
BROWSE Deer browse in the Ouachita forest in Oklahoma. J. Wildl. Manage. 32 658

BRUSH Brush control and the welfare of lesser prairie chickens in western O 188
BRUSH Effects on lesser prairie chickens of brush control in western Oklaho 189
BRUSH Brush control and pasture development in the Red Plains. Agron. J. 42 208
BRUSH New methods of brush control for more grass. J. Range Manage. 3: 46-5 210
BRUSH Rodent populations and their reactions to grazing intensities on sand 242
BRUSH Populations and range effects of rodents on the sand sagebrush grassl 457
BRUSH Spraying 2,4-D by airplane on sand sagebrush and other plants of the 469

BRYOPHYTA Bryophyta of southeast Oklahoma. I. A preliminary list with notes 672

BRYOPHYTES Bryophytes and pteridiophytes of some west central Oklahoma cany 405
BRYOPHYTES The bryophytes of Muskogee County, Oklahoma. Bryologist 38: 8-16 406
BRYOPHYTES Bryophytes of eastern Oklahoma. Proc. Okla. Acad. Sci. 9: 35-38. 670
BRYOPHYTES The bryophytes of Oklahoma: A preliminary study with notes on th 671

BUCKEYE Toxic plants of Oklahoma: Oak, black locust, Kentucky coffee-tree, 96
BUCKEYE The occurrence of persimmon and buckeye in the Wichita Mountains Wi 285

BUDGETS Calcium, magnesium, phosphorus, and potassium budgets of a tall gra 674

BUFFALO A preliminary study of buffalo wallows in the vicinity of Norman, O 45
BUFFALO The effect of early spring burning on vegetation in buffalo wallows 142
BUFFALO Effects of disturbance on diversity in mixed-grass prairie. Vegetat 144
BUFFALO An analysis of the vegetation of buffalo wallows in relation to env 579
BUFFALO The relationship of plant species heterogeneity to soil variation i 580
BUFFALO Relationships of vegetation and environment in buffalo wallows. Ame 581

BUFFALO Methods of reestablishing buffalo grass on cultivated land in the G 647

BURNING Differential response of woody and herbaceous species to summer and 9
BURNING The response of a central Oklahoma grassland to burning. Southw. Na 12
BURNING The effect of early spring burning on vegetation in buffalo wallows 142
BURNING Why prairies are treeless. Science N. S. 22: 55-56. {fire burning g 249
BURNING The effects of nitrogen fertilization on native vegetation under cl 250
BURNING The effects of one year's nitrogen fertilization on native vegetati 251
BURNING Burning and fertilization for range improvement in central Oklahoma 271
BURNING The effect of fire on the growth of Rhus glabra L. and Smilax bona- 332
BURNING Winter burning in central Oklahoma grassland. Ecology 38: 520-522. 370
BURNING A summary of fire and forage research on shinnery oak rangelands. P 468
BURNING The effect of disturbances in tall grass prairie sites on an index 503
BURNING Influence of a wildfire in the Wichita Mountains Wildlife Refuge. E 554
BURNING Some effects of winter burning on a moderately grazed pasture. Ecol 560
BURNING Causes of decreases in productivity in undisturbed tall grass prair 613
BURNING Ecosystem analysis of the tallgrass prairie: nitrogen cycle. Ecolog 636
BURNING Establishment of warm and cool season grass pastures on wooded site 664
BURNING The influence of burning on tillering and aboveground biomass parti 717

BURROWING Life history and habitat requirements of burrowing owls in wester 115

CACTACEAE Studies in the Cactaceae. Ph. D. Thesis, Univ. Okla., Norman. {fl 642

CALAMOVILFA Field investigations, McGee Creek Project, Oklahoma, Calamovilf 313
CALAMOVILFA Calamovilfa arcuata. Status report. Dept. of the Army, U. S. Ar 735
CALAMOVILFA Calamovilfa arcuata. Status report. Endangered Species Office, 745

CALLIRHOE Reproductive studies in the genus Callirhoe (Malvaceae). Ph. D. T 26
CALLIRHOE Field investigations, McGee Creek Project, Oklahoma, Calamovilfa 313
CALLIRHOE Callirhoe bushii. Status report. Dept. of the Army, U. S. Army Co 736

CALOPOGON Calopogon pulchellus from Oklahoma. Rhodora 36: 420-421. {flora} 259

CAMPANULACEAE Additions to the Oklahoma flora: Sphenoclea zeylanica (Camp 151

CANOPY Canopy structure of a tall-grass prairie. J. Range Manage. 27: 313-3 147

CANYON The vegetation of an eroding canyon in Canadian County, Oklahoma. M. 245
CANYON The vegetation of an actively eroding canyon in Canadian County, Okl 246
CANYON Bryophytes and pteridiophytes of some west central Oklahoma canyons. 405
CANYON The vegetation of the Caddo County Canyons, Oklahoma. Ecology 20: 1- 410
CANYON The microclimate of a relict stand of sugar maple in Devil's Canyon 600
CANYON Ecology and taxonomy of Water Canyon, Canadian County, Oklahoma. M. 722
CANYON The comparative ecology of two canyons and an upland area in west ce 837

CARBOHYDRASE Causes of decreases in residual carbohydrase activity in soil 612

CARBON A comparison of organic carbon, hydrogen ion concentration, and volu 368
CARBON Carbon dioxide exchange characteristics of some prairie grass seedli 630
CARBON Loss of ferulic acid from decomposing leaves of Celtis laevigata and 770

CAREX Carex latebracteata. Status report. Dept. of the Army, U. S. Army Cor 733
CAREX Carex latebracteata. Status report. Endangered Species Office, U. S. 750
CAREX A new species of Carex (Section Phyllostachyae) from Oklahoma. Rhodor 821

CARYA A limited-range provenance test of open-pollenated progeny of Carya i 300
CARYA The effect of water potential upon the geographic distribution of Car 318

CATALOG Annotated catalog of the ferns and flowering plants of Oklahoma. Ok 59

CATTLE Ornamental plants potentially hazardous to cattle. Bovine Pract. 18: 105

CELLULOSE Microbial decomposition of cellulose and native plant litter in a 447

CELTIS Allelopathic effects of Celtis laevigata Willd. as related to patter 417
CELTIS Allelopathic effects of Celtis laevigata. Bull. Torrey Bot. Club 98: 418
CELTIS Loss of ferulic acid from decomposing leaves of Celtis laevigata and 770

CEMBROID Studies in the cembroid pines. M. S. Thesis, Univ. Okla., Norman. 237

CEMENT-KILN The effect of cement-kiln dust on vegetation. M. S. Thesis, Uni 21

CENCHRUS A new sandbur from western Oklahoma: Cenchrus albertsonii. Amer. J 644

CERATOPHYLLACEAE Additions to the Oklahoma flora: Sphenoclea zeylanica 151

CERATOPHYLLUM Additions to the Oklahoma flora: Sphenoclea zeylanica 151

CERCIS Genecology of eastern redbud (Cercis canadensis). Ecology 63: 962-97 190

CHAETOPAPPA Systematics of Chaetopappa series Effusae. M. S. Thesis, Univ. 561

CHARA A population study of Chara zeylanica in Texas, Oklahoma, and New Mex 275

CHECKLIST Annotated checklist of the vascular plants of Little Sahara State 679
CHECKLIST Annotated checklist of the Compositae of Marshall County, Oklahom 721
CHECKLIST The genus Achillea. In: A synonimized checklist of the vascular f 777

CHEMICAL Chemical composition of certain aquatic plants. Bot. Gaz. 96: 186- 296
CHEMICAL Effects of chemical composition of soil on micronutrients and othe 298
CHEMICAL Chemical and biological analysis of a hazardous waste disposal sit 360
CHEMICAL The ecology of Honey Creek: Temporal patterns of the travertine pe 596
CHEMICAL Some roles of allelopathic compounds in plant communities. Biochem 609

CHISHOLM The Chisholm Trail in geographic perspective. Proc. Okla. Acad. Sc 761

CHLOROPHYCEAN A third species of Enteromorpha (a marine Chlorophycean) 148

CHLOROPHYLL Variation with stream order in species composition, diversity, 665

CHROMOSOME IOPB chromosome number reports. Taxon 18: 433. {flora} 182
CHROMOSOME Chromosome number of Phlox oklahomensis Wherry; IOPB 704
CHROMOSOME Chromosome numbers in Nama and Proboscidea; IOPB 780

CLASSIFICATION The biotic districts of Oklahoma. Amer. Midl. Natur. 20: 425 55
CLASSIFICATION The eastern boundary of the Great Plains in north central Ok 63
CLASSIFICATION Ordination and classification of mature bottomland forests i 140
CLASSIFICATION The animal and plant life in Oklahoma. Okla. Geol. Surv. Cir 166
CLASSIFICATION Plant associations as site predictors in the pine-hardwood t 187
CLASSIFICATION Ordination and classification of western oak forests in Okla 192
CLASSIFICATION A game type map of Oklahoma. Okla. Game and Fish Dept., Okla 194
CLASSIFICATION A survey of the game and furbearing animals of Oklahoma. Okl 195
CLASSIFICATION Identifying the prairie-forest transition using remote sensi 196
CLASSIFICATION The savannah concept and its use. Ecology 38: 435-442. {clas 200
CLASSIFICATION Classification of Oklahoma grasses. Ann. Okla. Acad. Sci. 6: 220
CLASSIFICATION Benchmark and key soils of Oklahoma. Okla. State Univ. Agr. 273
CLASSIFICATION The development of aquatic ecoregions in Oklahoma. Ph. D. Th 342
CLASSIFICATION The life-form concept applied to prairie chicken habitat in 354
CLASSIFICATION Identification and analysis of lesser and greater prai 356
CLASSIFICATION Potential natural vegetation of the conterminous United Stat 388
CLASSIFICATION A biogeographical boundary: The Tatschl Line. Trans. Kans. A 389

CLASSIFICATION Biotic type mapping of Oklahoma watersheds. Proc. Okla. Acad 521
CLASSIFICATION The savanna concept in Oklahoma. Ecology 43: 774-775. {all c 548
CLASSIFICATION The vegetation regions of the prairie province. Bot. Gaz. 25 585
CLASSIFICATION Habitat description and the life-form concept. Proc. Okla. A 707
CLASSIFICATION Primary succession on granite outcrops in southwestern Oklah 797
CLASSIFICATION Biogeographic regions of Texas and Oklahoma. Ecology 31: 427 833
CLASSIFICATION Oklahoma climagraphs and biotic regions. Proc. Okla. Acad. S 835
CLASSIFICATION Life zones and zone indicators in Oklahoma. Proc. Okla. Acad 841
CLASSIFICATION Natural areas for Oklahoma: important areas of natural signi 860

CLAY The claypan range site in northern Osage County, Oklahoma. J. Range Ma 305
CLAY A comparison of algal succession between farm ponds with sandstone and 856
CLAY A comparison of algal composition and seasonal succession between pond 857

CLAYPAN The claypan range site in northern Osage County, Oklahoma. J. Range 305

CLEARCUT Nutrient analysis of selected forbs on clearcut areas in southeast 594

CLIMAGRAPHS Oklahoma climagraphs and biotic regions. Proc. Okla. Acad. Sci. 835

CLIMATE Some factors influencing the frequency of bark lichens in north cen 5
CLIMATE Some factors influencing the frequency of bark lichens in north cen 6
CLIMATE Injury and death or recovery of trees in a prairie climate. Ecol. M 20
CLIMATE Growth and distribution of Platanus occidentalis in Oklahoma as inf 30
CLIMATE The climate of the central North American Grassland. Ann. Assoc. Am 68
CLIMATE Microclimates of three grassland plots in central Oklahoma. M. S. T 82
CLIMATE Weather patterns and ragweed production in Tulsa, Oklahoma. Ann. Al 87
CLIMATE Ragweed pollen and weather patterns in Tulsa, Oklahoma. Bull. Ecol. 88
CLIMATE An assessment of weather patterns and ragweed pollen in Tulsa over 89
CLIMATE Relation of climatic conditions, soil characteristics, and tree dev 294
CLIMATE Drought years in central Oklahoma from 1710 to 1959 calculated from 295
CLIMATE Correlation analysis of rainfall and annual ring index of central O 350
CLIMATE The microclimate of a relict stand of sugar maple in Devil's Canyon 600
CLIMATE The microclimate of sugar maple stands in Oklahoma. Ecology 43: 19- 601
CLIMATE Drought survival of native grass species in the central and souther 648
CLIMATE Effect of disturbance on microclimate and vegetation in tall grass 693
CLIMATE Climate and man. Yearbook of agriculture. U. S. Govt. Printing Offi 794
CLIMATE Oklahoma climagraphs and biotic regions. Proc. Okla. Acad. Sci. 5: 835

CLIMAX Inhibition of nitrification by climax ecosystems. Amer. J. Bot. 59: 615
CLIMAX Inhibition of nitrification by climax ecosystems. II. Additional evi 616
CLIMAX Inhibition of nitrification by climax ecosystems. III. Inhibitors ot 617
CLIMAX Allelopathic inhibition of weed invasion by Andropogon scoparius Mic 713

CLIPPING The effects of nitrogen fertilization on native vegetation under c 250
CLIPPING The effects of one year's nitrogen fertilization on native vegetat 251
CLIPPING Some effects of differential clipping on six native grasses and on 692

CLOVERS Toxic plants of Oklahoma: Clovers. J. Okla. Vet. Med. Assoc. 36: 30 203

CLUSTER Ordination and classification of mature bottomland forests in north 140
CLUSTER Description and dynamics of some western oak forests in Oklahoma. P 191
CLUSTER Ordination and classification of western oak forests in Oklahoma. A 192
CLUSTER Revegetation of unreclaimed coal strip mines in Oklahoma. II. Plant 256

CNIDOSCOLUS Pollination of Cnidoscolus texanus (Euphorbiaceae) in south cen 564

CO-DISPERSION Dispersion and co-dispersion in Oklahoma upland forests. Bull 678

COAL Revegetation of unreclaimed coal strip mines in Oklahoma. II. Plant co 256
COAL A comparison of forest vegetation on orphan strip mines and adjacent s 299
COAL Revegetation of unreclaimed coal strip-mines in Oklahoma. I. Vegetatio 348
COAL Revegetation of coal stripped land near Henryetta, Oklahoma. Proc. Okl 439
COAL Flora and fauna of Oklahoma abandoned mine lands. Okla. Biol. Surv., N 654
COAL Model fish and wildlife plan for a surface coal mine in Oklahoma. Okla 655
COAL Revegetation of strip mine spoils. M. S. Thesis, Okla. State Univ., St 702

COAST The Gulf Coastal Plain Ultisol and Alfisol ecosystem: surface geology 502
COAST Relationship of soil properties to site index of shortleaf pine and d 526
COAST Distribution, seasonal abundance, and hosts of the Gulf Coast tick in 662
COAST The relation of forest vegetation to soils and geology in the gulf co 729
COAST Shortleaf pine site index – Soils and plant associations on the coas 851

COASTAL The Gulf Coastal Plain Ultisol and Alfisol ecosystem: surface geolo 502
COASTAL Relationship of soil properties to site index of shortleaf pine and 526
COASTAL The relation of forest vegetation to soils and geology in the gulf 729
COASTAL Shortleaf pine site index – Soils and plant associations on the co 851

COCKLEBUR Toxic plants of Oklahoma: Cocklebur. J. Okla. Vet. Med. Assoc. 37 111

COFFEEWEEDS Toxic plants of Oklahoma: Coffeeweeds and sennas. J. Okla. 102

COLLECTIONS Collections towards a flora of the Territory of Arkansas. Trans 510
COLLECTIONS Plant collections representative of some typical plant communit 517
COLLECTIONS Plant collections representative of some typical communities in 518

COLORADO Survey of a wagon road from Fort Smith to the Colorado River. Chro 214
COLORADO The vegetation of the Mesa de Maya region of Colorado, New Mexico, 639

COMMUNITIES Water quality and phytoplankton communities in Lake Thunderbird 47
COMMUNITIES The relation of plant communities to soil erosion in Cleveland 65
COMMUNITIES Comparative studies of the animal communities of oak-hickory fo 76
COMMUNITIES Seasonal aspects in certain prairie and savannah communities. M 121
COMMUNITIES Fluctuations in biotic communities. III. Aspection in a ravine 124
COMMUNITIES Fluctuations in biotic communities. V. Aspection in mixed-grass 125
COMMUNITIES Grassland communities of the Wichita Mountains Wildlife Refuge. 162
COMMUNITIES A comparison of post oak-blackjack oak communities in two major 199
COMMUNITIES Plant communities and the potential for ecotypic differentiatio 254
COMMUNITIES Revegetation of unreclaimed coal strip mines in Oklahoma. II. P 256
COMMUNITIES Changes in species composition of grassland communities in resp 337
COMMUNITIES The quantitative phenology of two plant communities in Osage Co 355
COMMUNITIES Analysis of the diatom (Bacillariophyta) communities of the Nor 373
COMMUNITIES Cellular slime molds (Dictyosteliales) of selected plant commun 391
COMMUNITIES Relation between soil fungi and seed plants in three succession 437
COMMUNITIES Relation between soil fungi and seed plants in three succession 438
COMMUNITIES Plant collections representative of some typical plant communit 517
COMMUNITIES Plant collections representative of some typical communities in 518
COMMUNITIES Edaphic and topographic effects on forest communities in Payne 536
COMMUNITIES Plant communities of Oklahoma lakes. Ecology 34: 561-583. {aqua 543
COMMUNITIES A quantitative analysis of bottomland communities in south-cent 567
COMMUNITIES Some roles of allelopathic compounds in plant communities. Bioc 609
COMMUNITIES A comparison of the predominant soil microflora in eight vegeta 706
COMMUNITIES The grassland communities of the Baum Limestone in Johnston Cou 755

COMMUNITY Some observations on the number and size of samples necessary to 122

COMMUNITY The black-tailed prairie dog (Cynomys ludovicianus) as a member o 176
COMMUNITY Changes in community structure in Oklahoma old field succession. 364
COMMUNITY An analysis of an elm-ash floodplain community near Norman, Oklah 541
COMMUNITY A plant community dominated by vines. Proc. Okla. Acad. Sci. 45: 555
COMMUNITY Replacement of a population of Johnson grass by a vine-forb commu 556
COMMUNITY A study of the role of Rhus copallina in the community dynamics o 565
COMMUNITY Plant community structure. In: N. R. French (ed) Preliminary anal 627
COMMUNITY Primary succession on granite outcrops in southwestern Oklahoma. 797

COMPACTION A study of soil compaction in a tall grass prairie. M. S. Thesis 506

COMPETITIVE Competitive and allelopathic suppression of understory by Oklah 476
COMPETITIVE Competitive responses of Bouteloua curtipendula (Michx.) Torr. 624

COMPOSITAE Erigeron geiseri (Compositae) in Oklahoma. Southw. Natur. 24: 38 500
COMPOSITAE A study of the Compositae and Euphorbiaceae of Oklahoma and Clev 710
COMPOSITAE Annotated checklist of the Compositae of Marshall County, Oklaho 721
COMPOSITAE Origin and distribution of polyploid Achillea (Compositae) in we 774

COMPOSITION Chemical composition of certain aquatic plants. Bot. Gaz. 96: 1 296
COMPOSITION Effects of chemical composition of soil on micronutrients and o 298
COMPOSITION Forage production, vegetative composition, and plant vigor in r 304
COMPOSITION Effect of grazing intensity on plant composition, vigor, and pr 306
COMPOSITION Changes in species composition of grassland communities in resp 337
COMPOSITION Effects of moderate grazing on the composition and plant produc 369
COMPOSITION A study of the environment and floristic composition of an oak- 426
COMPOSITION Microbial decomposition of cellulose and native plant litter in 447
COMPOSITION The composition of a black willow forest in south central Oklah 547
COMPOSITION The composition of a post oak forest in south-central Oklahoma. 549
COMPOSITION Species composition and diversity of phytoplankton in the Grand 570
COMPOSITION Species composition and diversity of periphyton in the Grand Ri 571
COMPOSITION Effects of oat straw and native hay mulches on plant compositio 610
COMPOSITION Composition of a green ash forest near Norman, Oklahoma. Southw 620
COMPOSITION Ecosystem analysis of the tallgrass prairie: nitrogen cycle. Ec 636
COMPOSITION Variation with stream order in species composition, diversity, 665
COMPOSITION Microbial decomposition of ferulic acid in soil. J. Chem. Ecol. 771
COMPOSITION Studies in the composition and distribution of the Oklahoma flo 820
COMPOSITION A comparison of algal composition and seasonal succession betwe 857

COMPOUNDS Effects of bacterial inhibitors and certain related compounds on 513
COMPOUNDS Some roles of allelopathic compounds in plant communities. Bioche 609

CONSERVANCY The Nature Conservancy's first Oklahoma preserve. Proc. Okla. A 41

CONSERVATION Conservation ranching in the Oklahoma panhandle. J. Range Mana 31
CONSERVATION Bottomland hardwoods of eastern Oklahoma: A special study of t 72
CONSERVATION Revegetation and gully-control experiments in the Red Plains 209

CONTERMINOUS Potential natural vegetation of the conterminous United States 388

CORONASTRUM Coronastrum: a rare find. Trans. Amer. Micros. Soc. 96: 163. {a 568

CORRELATION Correlation analysis of rainfall and annual ring index of centr 350
CORRELATION Correlation between vegetation and geologic formations in Oklah 597

COTTON The comparative utility of plant life-form and other veg. chars. in 50
COTTON Toxic plants of Oklahoma: Cotton and cottonseed. J. Okla. Vet. Med. 202
COTTON The influence of habitat quality on density of cotton rat population 258

COTTONSEED Toxic plants of Oklahoma: Cotton and cottonseed. J. Okla. Vet. M 202

COTTONWOOD Economical harvesting systems for scattered stands of cottonwood 278
COTTONWOOD Cottonwood site evaluation in Oklahoma. M. S. Thesis, Okla. Stat 715

COUNTRY Journal of the expedition of the regiment of dragoons into Indian c 839

CROSS TIMBERS The Cross Timbers. Star Printery, Muskogee, OK. {upland fores 238
CROSS TIMBERS A floristic study of the woody vegetation of the North Americ 302
CROSS TIMBERS The detection and nature of functional underground connection 422
CROSS TIMBERS Forest type conversion on the "Cross Timbers" area in Oklahom 524

CROWN Growth of ten species of preferred deer browse plants under various c 132

CRYSTAL The algae of Crystal Lake, Cleveland County, Oklahoma. Amer. Midl. 397
CRYSTAL The algae of Crystal Lake, 1938-1978. M. S. Thesis, Univ. Okla., No 427
CRYSTAL Crystal Lake revisited. Proc. Okla. Acad. Sci. 61: 49-57. {SC Cleve 428

CULTIVATED Non-allergenic native and cultivated plants in Oklahoma: Landsca 400
CULTIVATED Data and observations concerning regrassing abandoned cultivated 475
CULTIVATED Methods of reestablishing buffalo grass on cultivated land in th 647

CUSCUTA Cuscuta attenuata. Status report. Endangered Species Office, U. S. 749

CYANOGENIC Toxic plants of Oklahoma: Cyanogenic plants. J. Okla. Vet. Med. 107
CYANOGENIC Plants of Oklahoma and Texas capable of producing cyanogenic poi 685

CYCLING Standing state and cycling of nitrogen in soil-vegetation component 61

CYPERACEAE Systematic studies in the Poaceae and Cyperaceae. Ph. D. Thesis, 73
CYPERACEAE A study of the angiosperms, exclusive of the Gramineae, Cyperace 153
CYPERACEAE Cyperus surinamensis (Cyperaceae) new to Arkansas, Kansas, and O 404
CYPERACEAE A systematic and ecological study of the Gramineae, Cyperaceae, 588
CYPERACEAE The spermatophyta of Cherokee County, Oklahoma (exclusive of Gra 802

CYPERUS Cyperus surinamensis (Cyperaceae) new to Arkansas, Kansas, and Okla 404

CYTOGEOGRAPHY Cytogeography and reproductive biology of Achillea millefoliu 577
CYTOGEOGRAPHY Cytogeography of Achillea millefolium in Oklahoma and 578

CYTOLOGICAL Cytological and morphological variation in Panicum virgatum L. 80

DAMS Changes in vegetation on stock pond dams near Norman, Oklahoma. M. S. 366
DAMS The vegetation of stock pond dams in central Oklahoma. Amer. Midl. Nat 371

DANTHONIEAE Studies in the Danthonieae (Poaceae). Ph. D. Thesis, Univ. Okla 854

DECIDUOUS Comparative water relations of four species of upland deciduous t 23
DECIDUOUS Deciduous forests of eastern North America. Blakiston, Philadelph 74

DECOMPOSITION Microbial decomposition of cellulose and native plant litter 447
DECOMPOSITION Ecosystem analysis of the tallgrass prairie: nitrogen cycle. 636
DECOMPOSITION Microbial decomposition of ferulic acid in soil. J. Chem. Eco 771

DENDROCHRONOLOGY Drought years in central Oklahoma from 1710 to 1959 295
DENDROCHRONOLOGY Correlation analysis of rainfall and annual ring index of 350

DENSITIES Growth of ten species of preferred deer browse plants under vario 132

DENSITY Some morphological and physiological effects of density and density 197
DENSITY The influence of habitat quality on density of cotton rat populatio 258

DENUDATION Effects of denudation on the productivity of grassland. Ecology 552

DENUDED Soil factors preventing revegetation of a denuded area near an aban 532

DESTRUCTION Plant succession following destruction of vegetation by toxins 827

DETECTION The detection and nature of functional underground connections be 422

DETERIORATION Range deterioration in south central Oklahoma. M. S. Thesis, 494

DEVELOPMENT Brush control and pasture development in the Red Plains. Agron. 208
DEVELOPMENT Evidence for the absence of ecotypic development in Andropogon 257
DEVELOPMENT A preliminary survey of important factors which affect tree dev 292
DEVELOPMENT Effect of silting on tree development in the flood plain of the 293
DEVELOPMENT Relation of climatic conditions, soil characteristics, and tree 294
DEVELOPMENT Developmental growth of four species of range grasses in north- 334
DEVELOPMENT The development of aquatic ecoregions in Oklahoma. Ph. D. Thesi 342
DEVELOPMENT The seeding, development, and utilization of native grasses in 485
DEVELOPMENT Growth and development of five species of range grasses in cent 598
DEVELOPMENT Plant development and sequence in artificial ponds with and wit 769
DEVELOPMENT Effects of excessive mulch on development, yield, and structure 832

DIATOM Some algae of Lake Carl Blackwell, Oklahoma, exclusive of diatoms. M 149
DIATOM Analysis of the diatom (Bacillariophyta) communities of the North Ca 373
DIATOM Additions to the diatom (Bacillariophyceae) flora of Oklahoma. Proc. 374
DIATOM Diatoms, including salt-water taxa, from southwestern Oklahoma. Proc 385
DIATOM Epiphytic diatoms as indicators of water quality in farm and experim 765
DIATOM Epiphytic diatoms in farm ponds and experimental ponds in Bryan Coun 767
DIATOM Epiphytic diatoms as indicators of water quality in farm and experim 768

DIFFERENTIATION Plant communities and the potential for ecotypic differenti 254

DIGESTED Restoration of eroded prairie with digested sludge. Proc. Okla. Ac 377

DIGITARIA Inhibitional effects of Digitaria sanguinalis and possible role i 533
DIGITARIA Inhibitional effects of Digitaria sanguinalis and possible role i 534

DINOFLAGELLATE Electrophoretic analysis of soluble enzymes in five freshwat 303
DINOFLAGELLATE Studies in the dinoflagellate genera, Peridinium and Peridin 321
DINOFLAGELLATE A survey of auxotrophy in five freshwater dinoflagellates (P 322

DISPERSAL Seed predation, seed dispersal, and disturbance in grasslands: a 143
DISPERSAL Seed dispersal and mineral nutrition in succession in abandoned f 611

DISPERSION Dispersion and co-dispersion in Oklahoma upland forests. Bull. T 678

DISTRESS Toxic plants of Oklahoma: Species causing acute respiratory distre 110

DISTRIBUTION Distribution and control of several woody plants in Texas and 25
DISTRIBUTION Growth and distribution of Platanus occidentalis in Oklahoma a 30
DISTRIBUTION New plant distribution records in eastern Oklahoma. Proc. Okla 53
DISTRIBUTION Silverleaf nightshade (Solanum eleagnifolium) origin, distribu 71
DISTRIBUTION Distribution and identification of woody plants of Oklahoma in 86
DISTRIBUTION The effect of salinity on the distribution of algae at the Sal 113
DISTRIBUTION Plant distribution in a central Oklahoma prairie. Proc. Okla. 123
DISTRIBUTION Insect distribution and seasonal succession in overgrazed and 157
DISTRIBUTION The distribution of Ephedra in Oklahoma. Proc. Okla. Acad. Sci 165
DISTRIBUTION Low intensity forest sampling through use of stem frequency di 167
DISTRIBUTION Silvics of forest trees of the United States. Agr. Handbook No 241
DISTRIBUTION A comparative study of the water relations of Quercus mariland 283
DISTRIBUTION Geographic distribution of two species of oaks in Oklahoma in 284

DISTRIBUTION The effect of water potential upon the geographic distribution 318
DISTRIBUTION Plankton production and species distribution in the limnologic 379
DISTRIBUTION Ecological factors affecting the distribution of woody vegetat 420
DISTRIBUTION An autecological study of Salix nigra, especially factors that 471
DISTRIBUTION Factors limiting the distribution of Salix nigra. Bull. Torrey 472
DISTRIBUTION The role of ecotypic variation in the distribution of the cent 473
DISTRIBUTION Distribution of the grasses (Gramineae) in relation to plate t 520
DISTRIBUTION Relationship of soil properties to site index of shortleaf pin 526
DISTRIBUTION The distribution of rodents in overgrazed and normal grassland 575
DISTRIBUTION Rodent distribution in overgrazed and normal grasslands. M. S. 576
DISTRIBUTION Factor analysis of tree distribution patterns in Oklahoma. Eco 656
DISTRIBUTION Shrub distribution patterns in Oklahoma with comparisons to th 657
DISTRIBUTION Distribution, seasonal abundance, and hosts of the Gulf Coast 662
DISTRIBUTION The ecology and behavior of the lone star tick. (Acarina: Ixod 663
DISTRIBUTION The bryophytes of Oklahoma: A preliminary study with notes on 671
DISTRIBUTION Distribution of legumes as correlated with surface geology and 673
DISTRIBUTION Distribution, habitat, and reproductive biology of Phlox oklah 703
DISTRIBUTION Distribution records of Oklahoma flowering plants. Proc. Okla. 714
DISTRIBUTION Biology and distribution of Hypoxis longii Fern. (Amaryllidace 753
DISTRIBUTION Comments on the distribution of some Oklahoma plants. Proc. Ok 756
DISTRIBUTION Stem frequency distribution and sampling of the hardwood compo 760
DISTRIBUTION Origin and distribution of polyploid Achillea (Compositae) in 774
DISTRIBUTION Distributional notes and some minor forms from Oklahoma. Rhodo 811
DISTRIBUTION Studies in the composition and distribution of the Oklahoma fl 820
DISTRIBUTION Atlas of the woody plants of Oklahoma. Okla. Biol. Surv., Norm 845

DISTURBANCE Interaction of disturbances in tallgrass prairie: a field exper 139
DISTURBANCE Seed predation, seed dispersal, and disturbance in grasslands: 143
DISTURBANCE Effects of disturbance on diversity in mixed-grass prairie. Veg 144
DISTURBANCE The effect of disturbances in tall grass prairie sites on an in 503
DISTURBANCE The relationship of plant species heterogeneity to soil variati 580
DISTURBANCE Effect of disturbance on microclimate and vegetation in tall gr 693

DIVERSITY Interaction of disturbances in tallgrass prairie: a field experim 139
DIVERSITY Effects of disturbance on diversity in mixed-grass prairie. Veget 144
DIVERSITY Spatial and temporal variation in productivity, species diversity 150
DIVERSITY The effect of disturbances in tall grass prairie sites on an inde 503
DIVERSITY Species composition and diversity of phytoplankton in the Grand R 570
DIVERSITY Species composition and diversity of periphyton in the Grand Rive 571
DIVERSITY Diversity in tree species in Oklahoma upland forests. Ecology 52: 632
DIVERSITY Variation with stream order in species composition, diversity, bi 665
DIVERSITY Primary succession on granite outcrops in southwestern Oklahoma. 797

DORMANCY Mechanism of seed dormancy in Ambrosia artemisiifolia. Ph. D. Diss 844

DRABA Draba aprica in Oklahoma. Rhodora 63: 223-225. {flora} 641
DRABA Draba aprica. Status report. Dept. of the Army, U. S. Army Corps of E 732

DRAGOONS Journal of the expedition of the regiment of dragoons into Indian 839

DRAINAGE Effect of silt on natural vegetation and drainage in the flood pla 297

DROSERA A preliminary report on the ecology of a Drosera meadow. Proc. Okla 39

DROUGHT Drought years in central Oklahoma from 1710 to 1959 calculated from 295
DROUGHT Drought survival of native grass species in the central and souther 648

DROWNING The drowning of trees in Lake Carl Blackwell. Proc. Okla. Acad. Sc 228

DUNE Vegetation of the Woods County, Oklahoma sand dunes. Ph. D. Thesis, Un 677
DUNE Annotated checklist of the vascular plants of Little Sahara State Park 679

DUST The effect of cement-kiln dust on vegetation. M. S. Thesis, Univ. Okla 21

DYNAMICS Description and dynamics of some western oak forests in Oklahoma. 191
DYNAMICS Biomass, annual net primary production and dynamics of six mineral 345
DYNAMICS Biomass, annual net primary production, and dynamics of six minera 351
DYNAMICS A study of the role of Rhus copallina in the community dynamics of 565
DYNAMICS The role of Rhus copallina in the dynamics of the forest-prairie e 566
DYNAMICS Herbage dynamics of a tall grass prairie: Osage, 1971. USIBP Grass 634
DYNAMICS Herbage dynamics of an Oklahoma tallgrass prairie: Osage 1972. USI 635
DYNAMICS The structure and function of ten western North American grassland 691

ECHINOCEREUS Biosystematic studies in the Echinocereus viridiflorus complex 398

ECOLOGICAL Ecological studies of plant inhibition by Solanum rostratum. Ph. 15
ECOLOGICAL Ecological interpretation for small mounds in landscapes of east 24
ECOLOGICAL Ecological relationships of the Bird Creek region, northeastern 54
ECOLOGICAL The possible ecological significance of inhibition of Euphorbia 75
ECOLOGICAL Overgrazing from an ecological point of view. Proc. Okla. Acad. 78
ECOLOGICAL Ecological studies in an eastern Oklahoma floodplain. Proc. Okla 93
ECOLOGICAL Some observations on the number and size of samples necessary to 122
ECOLOGICAL Ecological studies on the Canadian River floodplain in Cleveland 309
ECOLOGICAL Systematic study of the flowering plants in the vicinity of Norm 392
ECOLOGICAL An ecological study of the Southeastern Oklahoma Protective Unit 415
ECOLOGICAL Ecological factors affecting the distribution of woody vegetatio 420
ECOLOGICAL An autecological study of Salix nigra, especially factors that m 471
ECOLOGICAL The ecological basis of ornamental gardening in Oklahoma. M. S. 501
ECOLOGICAL Ecological study of western Payne County, Oklahoma land utilizat 535
ECOLOGICAL An outline for ecological life histories of herbaceous vascular 542
ECOLOGICAL A systematic and ecological study of the Gramineae, Cyperaceae, 588
ECOLOGICAL Abstracts of ecological literature pertaining to Oklahoma. M. S. 622
ECOLOGICAL The ecology and behavior of the lone star tick (Acarina: Ixodida 661
ECOLOGICAL Preliminary ecological study of the shinnery oak area of western 842

ECOLOGY Comparative pollination ecology of Pyrrhopappus geiseri and Pyrrhop 38
ECOLOGY A preliminary report on the ecology of a Drosera meadow. Proc. Okla 39
ECOLOGY Platt National Park: Environment and ecology. Univ. Okla. Press, No 42
ECOLOGY The pollination ecology of Solanum rostratum (Solanaceae). Amer. J. 69
ECOLOGY Ecology of the grassland biome. M. S. Thesis, Univ. Okla., Norman. 126
ECOLOGY Ecology of the nesting birds of the McCurtain Game Preserve, Oklaho 129
ECOLOGY Genecology of eastern redbud (Cercis canadensis). Ecology 63: 962-9 190
ECOLOGY Grasses and grasslands: systematics and ecology. Univ. Okla. Press, 218
ECOLOGY Observations on the effects of erosion on the ecology and maintenan 358
ECOLOGY The physiological ecology of upland and lowland Panicum virgatum. P 505
ECOLOGY The ecology of the western Oklahoma salt plains. Pub. Univ. Okla. B 519
ECOLOGY An outline for ecological life histories of herbaceous vascular hyd 542
ECOLOGY The ecology of Honey Creek: Temporal patterns of the travertine per 596
ECOLOGY Ecology of the scaled quail in the Oklahoma panhandle. Wildl. Monog 653
ECOLOGY The ecology and behavior of the lone star tick (Acarina: Ixodidae). 661
ECOLOGY The ecology and behavior of the lone star tick. (Acarina: Ixodidae) 663
ECOLOGY Ecology and taxonomy of Water Canyon, Canadian County, Oklahoma. 722
ECOLOGY The comparative ecology of two canyons and an upland area in west c 837

ECOLOGY The host plant-insect ecology of Pyrrhopappus (Asteraceae: Lactucea 861

ECOREGIONS The development of aquatic ecoregions in Oklahoma. Ph. D. Thesis 342

ECOSYSTEM Standing state and cycling of nitrogen in soil-vegetation compone 61
ECOSYSTEM The Gulf Coastal Plain Ultisol and Alfisol ecosystem: surface geo 502
ECOSYSTEM Inhibition of nitrification by climax ecosystems. Amer. J. Bot. 5 615
ECOSYSTEM Inhibition of nitrification by climax ecosystems. II. Additional 616
ECOSYSTEM Inhibition of nitrification by climax ecosystems. III. Inhibitors 617
ECOSYSTEM Ecosystem analysis of the tallgrass prairie: nitrogen cycle. Ecol 636
ECOSYSTEM The true prairie ecosystem. Hutchinson Ross Publ. Co., Stroudsbur 637
ECOSYSTEM The structure and function of ten western North American grasslan 689

ECOTONE Habitat preferences among representative wintering and breeding bir 353
ECOTONE A study of the role of Rhus copallina in the community dynamics of 565
ECOTONE The role of Rhus copallina in the dynamics of the forest-prairie ec 566

ECOTYPIC Plant communities and the potential for ecotypic differentiation o 254
ECOTYPIC Evidence for the absence of ecotypic development in Andropogon vir 257
ECOTYPIC The role of ecotypic variation in the distribution of the central 473

EDAPHIC Edaphic and topographic effects on forest communities in Payne Coun 536
EDAPHIC Effects of oat straw and native hay mulches on plant composition an 610

EFFICIENCIES Efficiencies of the point-center quarter and quadrat methods i 446
EFFICIENCIES The structure and function of ten North American grasslands. I 690

EFFUSAE Systematics of Chaetopappa series Effusae. M. S. Thesis, Univ. Okla 561

ELECTROPHORETIC Electrophoretic analysis of soluble enzymes in five freshwa 303

ELEMENTS Effects of chemical composition of soil on micronutrients and othe 298
ELEMENTS Biomass, annual net primary production and dynamics of six mineral 345
ELEMENTS Biomass, annual net primary production, and dynamics of six minera 351

ELM Vegetation response following winged elm (Ulmus alata Michx.) and oak (172
ELM Vegetational responses following winged elm and oak control in Oklahoma 173

ELM-ASH An analysis of an elm-ash floodplain community near Norman, Oklahom 541

ELYMUS Hybridization and introgression between Elymus canadensis and Elymus 498

EMERGENT Aquatic plants of Oklahoma: I. Submersed, floating-leaved, and sel 499

ENDANGERED Field study of the threatened and endangered plants of southeast 253
ENDANGERED Vegetation resources including endangered and threatened species 312
ENDANGERED A report on the threatened and endangered plant species in the B 631
ENDANGERED Herbaceous plants. In: Rare and endangered vertebrates and plant 700
ENDANGERED Veratrum woodii (Melanthium woodii). Status report. Endangered S 742
ENDANGERED Amorpha ouachitensis. Status report. Endangered Species Office, 743
ENDANGERED Hypoxis longii. Status report. Endangered Species Office, U. S. 744
ENDANGERED Calamovilfa arcuata. Status report. Endangered Species Office, U 745
ENDANGERED Monarda stipitatoglandulosa. Status report. Endangered Species O 746
ENDANGERED Phlox oklahomensis. Status report. Endangered Species Office, U. 747
ENDANGERED Lesquerella angustifolia. Status report. Endangered Species Offi 748
ENDANGERED Cuscuta attenuata. Status report. Endangered Species Office, U. 749
ENDANGERED Carex latebracteata. Status report. Endangered Species Office, U 750
ENDANGERED Streptanthus squamiformis. Status report. Endangered Species Off 751
ENDANGERED Phlox pilosa var. longipilosa. Status report. Endangered Species 752
ENDANGERED Field surveys and status evaluation on proposed and candidate en 781

ENDANGERED Endangered and threatened plants of Oklahoma. Dept. Botany-Micr 859

ENERGETICS The energetics of land use in three Oklahoma counties. Ph. D. Th 381

ENERGY Energy analyses of forestry practices in oak-pine forests and loblol 383
ENERGY Energy analysis of Oklahoma rangelands and improved pastures. J. Ran 384
ENERGY The structure and function of ten western North American grasslands. 689
ENERGY The structure and function of ten North American grasslands. III. Ne 690

ENTEROMORPHA A third species of Enteromorpha (a marine Chlorophycean) for 148
ENTEROMORPHA Enteromorpha, a marine alga in Oklahoma. Proc. Okla. Acad. S 569

ENVIRONMENT Platt National Park: Environment and ecology. Univ. Okla. Press 42
ENVIRONMENT Effects of various plant and environmental factors on post oak 159
ENVIRONMENT Final report on vegetation and microenvironments of Platt Natio 171
ENVIRONMENT An environmental study of the W. C. Austin Irrigation Project o 314
ENVIRONMENT Some vegetation-environment relationships in the upland forests 344
ENVIRONMENT Some vegetation-environment relationships in the upland forests 349
ENVIRONMENT A study of the environment and floristic composition of an oak- 426
ENVIRONMENT An analysis of the vegetation of buffalo wallows in relation to 579
ENVIRONMENT Relationships of vegetation and environment in buffalo wallows. 581
ENVIRONMENT Final environmental statement: Optima Lake, North Canadian Rive 782
ENVIRONMENT Final environmental statement: Hugo Lake, Kiamichi River, Oklah 783
ENVIRONMENT Final updated and revised environmental statement: Kaw Lake, Ar 784
ENVIRONMENT Notes on the growth and fruiting of Andropogon scoparius, with 806
ENVIRONMENT Environmental parameters and site identification for potential 853

ENZYMES Electrophoretic analysis of soluble enzymes in five freshwater dino 303
ENZYMES Soil enzymes in relation to old field succession: amylase, cellulas 531

EPHEDRA The distribution of Ephedra in Oklahoma. Proc. Okla. Acad. Sci. 17: 165

EPHEMERAL Sediment-precipitation-runoff relationships for the three ephemer 638

EPIPHYTIC Epiphytic diatoms as indicators of water quality in farm and expe 765
EPIPHYTIC Epiphytic diatoms in farm ponds and experimental ponds in Bryan C 767
EPIPHYTIC Epiphytic diatoms as indicators of water quality in farm and expe 768

EQUITABILITY The effect of disturbances in tall grass prairie sites on an i 503

ERIGERON Erigeron geiseri (Compositae) in Oklahoma. Southw. Natur. 24: 386- 500

ERIOGONUM A taxonomic study of the subgenus Pterogonum of the genus Eriogon 311

ERODED Investigations in erosion control and reclamation of eroded land at 174
ERODED Seed hay for regrassing eroded land. Proc. Okla. Acad. Sci. 25: 43-4 206
ERODED Restoration of eroded prairie with digested sludge. Proc. Okla. Acad 377
ERODED Revegetation of eroded lands in the High Plains of Texas and Oklahom 508
ERODED Biotic and physiographic succession on abandoned eroded farmland. Ph 695
ERODED Biotic and physiographic succession on abandoned eroded farmland. Ec 697

ERODING The vegetation of an eroding canyon in Canadian County, Oklahoma. M 245
ERODING The vegetation of an actively eroding canyon in Canadian County, Ok 246

EROSION The relation of plant communities to soil erosion in Cleveland Coun 65
EROSION Algae as pioneers in plant succession and their importance in erosi 66
EROSION Investigations in erosion control and reclamation of eroded land at 174
EROSION Observations on the effects of erosion on the ecology and maintenan 358
EROSION Restoration of eroded prairie with digested sludge. Proc. Okla. Aca 377
EROSION Wildlife and roadside erosion in central Oklahoma. Proc. Okla. Acad 694

EROSION The effect of overgrazing and erosion upon the biota of the mixed g 696

ESTABLISHMENT Grass stand establishment on rough, rocky, wooded lands in ea 423
ESTABLISHMENT Establishment and management of tame and native grass pasture 587
ESTABLISHMENT Establishment of warm and cool season grass pastures on woode 664

EUPHORBIA The possible ecological significance of inhibition of Euphorbia s 75
EUPHORBIA Pollination of Cnidoscolus texanus (Euphorbiaceae) in south centr 564
EUPHORBIA A study of the Compositae and Euphorbiaceae of Oklahoma and Cleve 710
EUPHORBIA A new species of Euphorbia from Oklahoma. Rhodora 50: 63-64. {flo 812

EUPHORBIACEAE Pollination of Cnidoscolus texanus (Euphorbiaceae) in south c 564
EUPHORBIACEAE A study of the Compositae and Euphorbiaceae of Oklahoma an 710

EURASIAN Eurasian water milfoil in Oklahoma. Proc. Okla. Acad. Sci. 58: 4-5 156
EURASIAN Eurasian watermilfoil – A new menace to Oklahoma waters. Proc. Ok 341

EUTHAMIA New species, new combinations, and notes on the goldenrods (Eutham 725

EXCHANGE Carbon dioxide exchange characteristics of some prairie grass seed 630

EXCLUSION Possible mechanisms of the exclusion of Johnson grass by tall gra 660

EXPEDITION Commerce of the prairies: or, The journal of a Santa Fe trader, 274
EXPEDITION Account of an expedition from Pittsburg to the Rocky Mountains, 340
EXPEDITION [reports of railroad route expeditions]. 33rd. Cong., 2nd.Sess., 787
EXPEDITION Journal of the expedition of the regiment of dragoons into India 839

EXPLORATION Exploration of the Red River of Louisiana. Senate Ex. Doc. No. 443
EXPLORATION Reports of explorations and surveys ... [long title – deals wi 788
EXPLORATION [long title concerning exploration of railroad routes]. 33rd. C 789

EXPLORATORY An exploratory study on soil algae. Ecology 38: 309-313. {SC gr 848

EXPOSURE Vegetation in relation to slope exposure and geology in the Arbuck 335

FARM Evaluation of Payne County, Oklahoma farmlands and vegetation patterns 177
FARM Natural revegetation of abandoned farmland in the central and southern 650
FARM Biotic and physiographic succession on abandoned eroded farmland. Ph. 695
FARM Biotic and physiographic succession on abandoned eroded farmland. Ecol 697
FARM Epiphytic diatoms as indicators of water quality in farm and experimen 765
FARM Survey of algae in farm ponds in Bryan County, Oklahoma. Southw. Natur 766
FARM Epiphytic diatoms in farm ponds and experimental ponds in Bryan County 767
FARM Epiphytic diatoms as indicators of water quality in farm and experimen 768
FARM A comparison of algal succession between farm ponds with sandstone and 856

FARMLAND Evaluation of Payne County, Oklahoma farmlands and vegetation patt 177
FARMLAND Natural revegetation of abandoned farmland in the central and sout 650
FARMLAND Biotic and physiographic succession on abandoned eroded farmland. 695
FARMLAND Biotic and physiographic succession on abandoned eroded farmland. 697

FAUNA Flora and fauna of Oklahoma abandoned mine lands. Okla. Biol. Surv., 654

FEEDING Habitat used by lesser prairie chickens for feeding related to seas 357

FENCING Effects of fencing and plowing on plant succession in a revegetatin 557

FERN Annotated catalog of the ferns and flowering plants of Oklahoma. Okla. 59
FERN Toxic plants of Oklahoma: Ferns and horsetails. J. Okla. Vet. Med. Ass 106
FERN Ferns of Oklahoma. Amer. Midl. Natur. 12: 91-112. {fern flora} 112
FERN The ferns of Oklahoma (revised). Okla. A. & M Exp. Sta. Circ. No. C-80 230

FERN A preliminary list of the ferns and seed plants of Oklahoma. Univ. Okl 343

FERTILIZATION The effects of nitrogen fertilization on native vegetation un 250
FERTILIZATION The effects of one year's nitrogen fertilization on native ve 251
FERTILIZATION Burning and fertilization for range improvement in central Ok 271
FERTILIZATION Effect of fertilization on native grass pastures in Oklahoma. 331

FERULIC Loss of ferulic acid from decomposing leaves of Celtis laevigata an 770
FERULIC Microbial decomposition of ferulic acid in soil. J. Chem. Ecol. 1: 771

FEVER Hay fever plants in Oklahoma. Ann. Allerg. 45: 26-32. {allergy} 399
FEVER Non-allergenic native and cultivated plants in Oklahoma: Landscaping 400

FIRE Differential response of woody and herbaceous species to summer and wi 9
FIRE The response of a central Oklahoma grassland to burning. Southw. Natur 12
FIRE Interaction of disturbances in tallgrass prairie: a field experiment. 139
FIRE Effects of disturbance on diversity in mixed-grass prairie. Vegetatio 144
FIRE Why prairies are treeless. Science N. S. 22: 55-56. {fire burning gras 249
FIRE The effect of fire on the growth of Rhus glabra L. and Smilax bona-nox 332
FIRE A quantitative comparison between an oak forest and an oak savannah in 352
FIRE Winter burning in central Oklahoma grassland. Ecology 38: 520-522. {SC 370
FIRE A summary of fire and forage research on shinnery oak rangelands. Proc 468
FIRE The effect of disturbances in tall grass prairie sites on an index of 503
FIRE Influence of a wildfire in the Wichita Mountains Wildlife Refuge. Ecol 554
FIRE Some effects of winter burning on a moderately grazed pasture. Ecology 560
FIRE Causes of decreases in productivity in undisturbed tall grass prairie. 613
FIRE Ecosystem analysis of the tallgrass prairie: nitrogen cycle. Ecology 6 636
FIRE The influence of burning on tillering and aboveground biomass partitio 717

FLINT HILLS Vegetation-soil relationships in Flint Hills bluestem pastures. 27

FLOOD Bottomland hardwoods of eastern Oklahoma: A special study of their st 72
FLOOD Ecological studies in an eastern Oklahoma floodplain. Proc. Okla. Aca 93
FLOOD Reproduction of woody plants in the absence of grazing. Proc. Okla. A 119
FLOOD The drowning of trees in Lake Carl Blackwell. Proc. Okla. Acad. Sci. 228
FLOOD Silting and forest succession on Deep Fork in southwestern Creek Coun 229
FLOOD Effect of silting on tree development in the flood plain of the North 293
FLOOD Effect of silt on natural vegetation and drainage in the flood plain 297
FLOOD Ecological studies on the Canadian River floodplain in Cleveland Coun 309
FLOOD The effects of periodic flooding on upland forest succession. M. S. T 387
FLOOD The mammals of Harmon County, Oklahoma. Proc. Okla. Acad. Sci. 49: 42 444
FLOOD Factors limiting the distribution of Salix nigra. Bull. Torrey Bot. C 472
FLOOD An analysis of an elm-ash floodplain community near Norman, Oklahoma. 541
FLOOD The composition of a black willow forest in south central Oklahoma. P 547
FLOOD A quantitative analysis of bottomland communities in south-central Ok 567
FLOOD Bottomland forests of north-central Oklahoma. Ecology 46: 708-714. {N 605
FLOOD Some variations of the fruit of Quercus macrocarpa. M. S. Thesis, Okl 708
FLOOD A study of revegetation in the floodplain of the South Canadian River 807
FLOOD The vegetation of the lower levels of the South Canadian River in cen 808

FLORA Reproductive studies in the genus Callirhoe (Malvaceae). Ph. D. Thesi 26
FLORA The range of Cypripedium kentuckiense. Amer. Orchid Soc. Bull. 54(10) 32
FLORA Flora of Kiowa County, Oklahoma. M. S. Thesis, Okla. A & M Coll., Sti 34
FLORA A floristic study of the vascular plants of the Gypsum Hills and Redb 35
FLORA Floristic components of the Gypsum Hills and Redbed Plains of southwe 36
FLORA Taxonomic studies in the Verbena stricta complex. Ph. D. Thesis, Univ 37

FLORA A preliminary survey of the vascular plants of Pottawatomie County, O 43
FLORA A preliminary list of the vascular plants of Pottawatomie County, Okl 44
FLORA Lilium michiganense Farwell (Liliaceae) and Gentiana alba Muhl. (Gent 46
FLORA New plant distribution records in eastern Oklahoma. Proc. Okla. Acad. 53
FLORA Annotated catalog of the ferns and flowering plants of Oklahoma. Okla 59
FLORA Native Oklahoma plants. Okla. Agric. Exp. Sta. Bull. No. 45. {flora a 60
FLORA Biosystematics of Artemisia carruthii and Artemisia mexicana. M. S. T 62
FLORA The pollination ecology of Solanum rostratum (Solanaceae). Amer. J. B 69
FLORA The tallgrass prairie orchids Platanthera leucophaea (Nutt.) Lindl. a 70
FLORA Silverleaf nightshade (Solanum eleagnifolium) origin, distribution, a 71
FLORA Systematic studies in the Poaceae and Cyperaceae. Ph. D. Thesis, Univ 73
FLORA Analysis of morphological variation in Schizachyrium scoparium (littl 77
FLORA Cytological and morphological variation in Panicum virgatum L. M. S. 80
FLORA Distribution and identification of woody plants of Oklahoma in the wi 86
FLORA Some additions to Eskew's list of flowering plants of the Wichita Mou 90
FLORA Vascular plants of Greer County, Oklahoma. M. S. Thesis, Univ. Okla., 92
FLORA Ferns of Oklahoma. Amer. Midl. Natur. 12: 91-112. {fern flora} 112
FLORA A list of some of the most interesting species of plants collected in 114
FLORA A preliminary survey of the wild flowering plants of Kay County. M. S 116
FLORA Observations on the native plants of Oklahoma Territory and adjacent 118
FLORA A study of the flowering plants of Tulsa County, Oklahoma, exclusive 133
FLORA The arborescent flora of Oklahoma. M. S. Thesis, Univ. Wyo., Laramie. 134
FLORA Sida in Oklahoma. Rhodora 50: 99. {flora} 136
FLORA Additions to the Oklahoma flora: Sphenoclea zeylanica (Campanulaceae) 151
FLORA A study of the angiosperms, exclusive of the Gramineae, Cyperaceae, a 153
FLORA The distribution of Ephedra in Oklahoma. Proc. Okla. Acad. Sci. 17: 6 165
FLORA The animal and plant life in Oklahoma. Okla. Geol. Surv. Circ. No. 6. 166
FLORA A preliminary survey of the flora of the Arbuckle Mountains. M. S. Th 168
FLORA A preliminary survey of the flora of the Arbuckle Mountains. Texas J. 169
FLORA IOPB chromosome number reports. Taxon 18: 433. {flora} 182
FLORA Relationships of two isolated groups of sugar maple in central Oklaho 183
FLORA Relationships of two isolated groups of sugar maples (Acer saccharum 184
FLORA An annotated plant list for Adams Ranch. M. S. Thesis, Ft. Hays Kansa 198
FLORA A taxonomic investigation of the Polygonum pensylvanicum L. (Polygona 201
FLORA The genetic relationship between Nama hispidum and Nama stevensii. M. 215
FLORA The flowering plants of the Wichita National Forest. M. S. Thesis, Un 216
FLORA The flowering plants of the Wichita Mountains Wildlife Refuge. Amer. 217
FLORA Classification of Oklahoma grasses. Ann. Okla. Acad. Sci. 6: 141-161. 220
FLORA A new species of grama grass. Bot. Gaz. 91: 103-105. {Comanche SW flo 225
FLORA Grasses of Oklahoma. Okla. A & M Coll. Tech. Bull. No. 3. {flora all} 226
FLORA A new Triodia from Oklahoma. Rhodora 40: 243-244. {flora} 227
FLORA The ferns of Oklahoma (revised). Okla. A. & M Exp. Sta. Circ. No. C-8 230
FLORA A flora of Oklahoma by T. R. Stemen and W. S. Myers – A review. Rhod 232
FLORA Gray's manual of botany, 8th edition. American Book Co., New York. {N 233
FLORA A monographic study of the genus Aster in Oklahoma. M. S. Thesis, Uni 236
FLORA Studies in the cembroid pines. M. S. Thesis, Univ. Okla., Norman. {PH 237
FLORA Silvics of forest trees of the United States. Agr. Handbook No. 271, 241
FLORA A preliminary list of the plants of the Arbuckle Mountains. B. A. The 244
FLORA Relationships of sugar maples (Acer saccharum and A. grandidentatum) 252
FLORA Field study of the threatened and endangered plants of southeastern O 253
FLORA Calopogon pulchellus from Oklahoma. Rhodora 36: 420-421. {flora} 259
FLORA Glottidium vesicarium in Oklahoma. Rhodora 36: 118-119. {flora} 260
FLORA Further notes on Oklahoma plants. Rhodora 38: 239-240. {flora} 261

FLORA A new species of Streptanthus. Rhodora 58: 354-355. {flora} 262
FLORA Spring flora of central Oklahoma. Univ. Okla. Dup. Serv., Norman. {SC 263
FLORA The Oklahoma botanical travels of G. W. Stevens. Proc. Okla. Acad. Sc 264
FLORA The botanical itineraries of A. H. Van Vleet. Chron. Okla. 56: 322-33 265
FLORA Notes on trees, shrubs, and vines in the Cherokee Nation. Trans. Kans 266
FLORA A study of the genus Psoralea in Oklahoma. M. S. Thesis, Univ. Okla., 270
FLORA A biosystematic study of the relationship of Nama hispidum and Nama s 276
FLORA A taxonomic study of the genus Juniperus in Oklahoma. M. S. Thesis, U 280
FLORA A nature study outline on trees and shrubs for the first six grades. 290
FLORA The identification of some of the more common native Oklahoma grasses 301
FLORA A floristic study of the woody vegetation of the North American Cross 302
FLORA Lichens from the state of Oklahoma. Pap. Mich. Acad. Sci. Arts. Lett. 307
FLORA A taxonomic study of the subgenus Pterogonum of the genus Eriogonum (311
FLORA List of plants collected by C. S. Sheldon and M. A. Carlton in the In 323
FLORA Oklahoma spring flowers in autumn. Proc. Okla. Acad. Sci. 17: 60-62. 325
FLORA Isolation of plants in the Oklahoma flora. Proc. Okla. Acad. Sci. 18: 326
FLORA Notes from the herbarium of the University of Oklahoma. Rhodora 40: 4 327
FLORA The floristic affinities of the Arbuckle Mountains in Oklahoma. Amer. 328
FLORA Notes from the Bebb Herbarium of the University of Oklahoma II. Rhodo 329
FLORA Notes on Oklahoma plants. Rhodora 45: 113-119. {flora} 330
FLORA A preliminary list of the ferns and seed plants of Oklahoma. Univ. Ok 343
FLORA The lichens of north central Oklahoma. Ph. D. Diss., Okla. State Univ 365
FLORA Additions to the diatom (Bacillariophyceae) flora of Oklahoma. Proc. 374
FLORA A preliminary survey of the flowering plants of Cleveland County. M. 380
FLORA Observations of the flora of Delaware County, Oklahoma. Proc. Okla. A 390
FLORA Systematic study of the flowering plants in the vicinity of Norman wi 392
FLORA The genus Tilia in Oklahoma. Proc. Okla. Acad. Sci. 50: 104-105. {SE 394
FLORA Galium pedemontanum (Bell.) All. (Rubiaceae) reported from Oklahoma. 395
FLORA The genus Galium (Rubiaceae) in the southeastern United States. Ph. D 396
FLORA Biosystematic studies in the Echinocereus viridiflorus complex. Ph. D 398
FLORA Panax quinquefolium (Araliaceae) in southeastern Oklahoma. Rhodora 87 403
FLORA Cyperus surinamensis (Cyperaceae) new to Arkansas, Kansas, and Oklaho 404
FLORA Bryophytes and pteridiophytes of some west central Oklahoma canyons. 405
FLORA The bryophytes of Muskogee County, Oklahoma. Bryologist 38: 8-16. {NE 406
FLORA The liverworts of Oklahoma. Bryologist 39: 25-34. {flora all} 407
FLORA Flora of Muskogee County, Oklahoma. Amer. Midl. Natur. 19: 369-389. { 408
FLORA Acer grandidentatum in Oklahoma. Rhodora 46: 445-450. {SW Canadian Ca 411
FLORA Botanical notes from southeastern Oklahoma. Proc. Okla. Acad. Sci. 31 412
FLORA Forest trees of Oklahoma, 12th. edition. Okla. For. Div., State Dept. 414
FLORA Collecting in Oklahoma. Bryologist 22: 21-22. {flora} 425
FLORA Vascular plants new for Oklahoma, Kansas, and Nebraska. Rhodora 73: 3 429
FLORA The native orchids of the prairies and plains region of North America 430
FLORA Listera australis Lindley (Orchidaceae) new for Oklahoma. Southw. Nat 431
FLORA New native orchids for Oklahoma and Kansas. Proc. Okla. Acad. Sci. 64 432
FLORA Glaucium flavum Crantz (Papaveraceae) and Silybum marianum (L.) Gaert 433
FLORA Orchids and other new and interesting plants from Oklahoma. Pub. No. 434
FLORA The genus Baccharis in the southwestern states of Oklahoma, Texas, an 435
FLORA Notes on rare Texas and Oklahoma plants. Sida 9: 76-86. {flora} 436
FLORA Seasonal algal flora on the campus of Catholic College, Guthrie, Okla 440
FLORA A numerical taxonomic study of Monarda (Labiatae) in southern Oklahom 441
FLORA Pollination biology of Polygala alba Nutt. (Polygalaceae). Ph. D. The 445
FLORA Preliminary list of flowering plants of Payne County, Oklahoma. M. S. 448
FLORA Reproductive biology of Commelina erecta (Commelinaceae). pp. 57-66, 449

FLORA Roadside flowers of Oklahoma. C & J Printing Co., Lawton, OK. {flora 450
FLORA Roadside flowers of Oklahoma. Vol. 2. C & J Printing Co., Lawton, Ok. 451
FLORA Roadside wild fruits of Oklahoma. Univ. Okla. Press, Norman. {flora} 452
FLORA Roadside trees and shrubs of Oklahoma. Univ. Okla. Press, Norman. {fl 453
FLORA A taxonomic study of the Umbelliferae in Oklahoma. M. S. Thesis, Univ 454
FLORA A taxonomic study of the vascular plants indigenous to, or naturalize 455
FLORA Vascular plants of Pontotoc County, Oklahoma. Amer. Midl. Natur. 59: 456
FLORA A floristic study of the native or naturalized vegetation of Washingt 459
FLORA A floristic study of Washington county, Oklahoma. Proc. Okla. Acad. S 460
FLORA An investigation of the soil microflora of two grassland plots. Proc. 462
FLORA New and interesting plants from the Great Plains. Rhodora 74: 378-388 463
FLORA Atlas of the flora of the Great Plains. Iowa State Univ. Press, Ames. 464
FLORA Flora of the Great Plains. Univ. Kansas Press, Lawrence. {flora tallg 465
FLORA Vascular plants of southeastern Oklahoma from the Sans Bois to the Ki 478
FLORA The spermatophyta of Custer County, Oklahoma. M. S. Thesis, Univ. Okl 479
FLORA A preliminary report on the flora of the Wichita Mountains. M. S. The 492
FLORA Aquatic plants of Oklahoma: I. Submersed, floating-leaved, and select 499
FLORA Erigeron geiseri (Compositae) in Oklahoma. Southw. Natur. 24: 386-387 500
FLORA Collections towards a flora of the Territory of Arkansas. Trans. Amer 510
FLORA Observations upon some Oklahoma plants. Bull. Torrey Bot. Club. 22: 3 512
FLORA Distribution of the grasses (Gramineae) in relation to plate tectonic 520
FLORA The ligneous flora of Rich Mountain, Arkansas and Oklahoma. J. Arnold 527
FLORA Leaves from a collector's notebook. J. Arnold Arb. 13:417-437. {flora 529
FLORA Notes on some plants of Oklahoma. J. Arnold Arb. 15: 127-134. {flora} 530
FLORA Systematics of Chaetopappa series Effusae. M. S. Thesis, Univ. Okla., 561
FLORA The reproductive biology of Proboscidea louisianica (Martyniaceae). R 573
FLORA The reproductive biology of Proboscidea louisianica (Martyniaceae). M. 574
FLORA Cytogeography and reproductive biology of Achillea millefolium in Okl 577
FLORA Cytogeography of Achillea millefolium in Oklahoma and adjacent states 578
FLORA A systematic and ecological study of the Gramineae, Cyperaceae, and J 588
FLORA A biosystematic study of Artemisia caruthii. Ph. D. Dissertation, Uni 591
FLORA A numerical analysis of the morphological variation in the genus Alet 623
FLORA A report on the threatened and endangered plant species in the BLM so 631
FLORA Draba aprica in Oklahoma. Rhodora 63: 223-225. {flora} 641
FLORA Studies in the Cactaceae. Ph. D. Thesis, Univ. Okla., Norman. {flora} 642
FLORA A new sandbur from western Oklahoma: Cenchrus albertsonii. Amer. J. B 644
FLORA A list of the flowering plants of Norman and vicinity. B. A. Thesis, 645
FLORA Flora and fauna of Oklahoma abandoned mine lands. Okla. Biol. Surv., 654
FLORA The trees and shrubs of Oklahoma. Okla. Geol. Surv. Circ. 4: 7-41. {f 666
FLORA Bryophytes of eastern Oklahoma. Proc. Okla. Acad. Sci. 9: 35-38. {NE 670
FLORA The bryophytes of Oklahoma: A preliminary study with notes on their e 671
FLORA Bryophyta of southeast Oklahoma. I. A preliminary list with notes. Br 672
FLORA Biosystematic study of the Spiranthes cernua complex. Bull. 448, New 680
FLORA The prairie fringed orchids: A pollinator-isolated species pair. Rhod 681
FLORA A partial list of the angiosperms of Garfield County. Proc. Okla. Aca 682
FLORA The woody plants of Oklahoma. Ph. D. Thesis, Univ. Calif., Berkeley. 683
FLORA Hedyotis ouachitana (Rubiaceae): a new species from the Ouachita Moun 698
FLORA Herbaceous plants. In: Rare and endangered vertebrates and plants of 700
FLORA Distribution, habitat, and reproductive biology of Phlox oklahomensis 703
FLORA A comparison of the predominant soil microflora in eight vegetative c 706
FLORA Oklahoma flora. Harlow Pub. Co., Okla. City. {all} 709
FLORA A study of the Compositae and Euphorbiaceae of Oklahoma and Cleveland 710
FLORA The flora of Oklahoma. unpub. ms. in Harvard Univ. Library, Cambridge 711

FLORA Plant life in Oklahoma. Okla. Geol. Surv. Bull. No. 27: 215-246. {flo 712
FLORA Distribution records of Oklahoma flowering plants. Proc. Okla. Acad. 714
FLORA Systematic studies in Physalis (Solanaceae). Ph. D. Thesis, Univ. Okl 716
FLORA A continuation of the investigation of the soil microflora of two gra 718
FLORA Annotated checklist of the Compositae of Marshall County, Oklahoma. P 721
FLORA Ecology and taxonomy of Water Canyon, Canadian County, Oklahoma. M. S 722
FLORA Comments on the flora of Bryan County. Proc. Okla. Acad. Sci. 44: 6-1 723
FLORA Podophyllum peltatum f. deamii from Bryan County, Oklahoma. Rhodora 6 724
FLORA New species, new combinations, and notes on the goldenrods (Euthamia 725
FLORA Solidago (Asteraceae) in Oklahoma and Texas. Sida 10: 223-251. {flora 726
FLORA Additions to the vascular flora of Oklahoma. Rhodora 67: 191-193. {} 737
FLORA The vascular flora of Oklahoma – additions and comments. Rhodora 71: 738
FLORA Additions to the vascular flora of Oklahoma – IV. Sida 12: 233-237. 739
FLORA Additions to the vascular flora of Oklahoma. Sida 7: 361-368. {} 740
FLORA Plants new to Oklahoma, Arkansas, and Texas. Sida 9: 25-28. {flora} 741
FLORA Biology and distribution of Hypoxis longii Fern. (Amaryllidaceae). Bu 753
FLORA Phlox pilosa L. var. longipilosa – A new combination. Sida 9: 183. { 754
FLORA Comments on the distribution of some Oklahoma plants. Proc. Okla. Aca 756
FLORA Allium ampeloprasum (Liliaceae) and Trifolium vesiculosum (Leguminose 757
FLORA Floral biology of Proboscidea louisianica (Martyniaceae). Rhodora 78: 758
FLORA Juvenile leaves in Oklahoma Marsilea (Marsileaceae). Sida 7: 218-219. 759
FLORA Some account of a collection of plants made during a journey to and f 764
FLORA Origin and distribution of polyploid Achillea (Compositae) in western 774
FLORA Identification and mapping of the extant flora at the Deer Creek arch 776
FLORA The genus Achillea. In: A synonimized checklist of the vascular flora 777
FLORA Nomenclature of the genus Achillea. In: National list of scientific n 778
FLORA A biosystematic study of the relationship of Nama hispidum and Nama s 779
FLORA Chromosome numbers in Nama and Proboscidea; IOPB Chromosome Repo 780
FLORA Field surveys and status evaluation on proposed and candidate endange 781
FLORA A biological inventory of the Fort Gibson Lake area. U. S. Dept. of t 796
FLORA Solidago flaccidifolia Small (Asteraceae) new to Oklahoma and other g 798
FLORA Plants of Oklahoma. Dept. Geol. and Nat. Hist.,2nd.bienn.rpt.,1901-19 799
FLORA Systematics of Aristida dichotoma, basiramea, and curtisii (Poaceae). 800
FLORA The spermatophyta of Cherokee County, Oklahoma (exclusive of Graminea 802
FLORA Additions to the Oklahoma flora from the Oklahoma Ozarks. Proc. Okla. 803
FLORA Vascular plants of the Oklahoma Ozarks. Ph. D. Diss., Okla. State Uni 804
FLORA Additions to the Oklahoma flora from Cherokee County. Proc. Okla. Aca 805
FLORA Interesting plants of Oklahoma. Rhodora 42: 499-502. {flora} 809
FLORA The spermatophyta of Oklahoma County, Oklahoma, exclusive of the gras 810
FLORA Distributional notes and some minor forms from Oklahoma. Rhodora 50: 811
FLORA A new species of Euphorbia from Oklahoma. Rhodora 50: 63-64. {flora} 812
FLORA Some results of a summer's botanizing in Oklahoma. Rhodora 51: 18. {f 813
FLORA A new species of Sarcostemma from Oklahoma. Rhodora 51: 58-59. {flora 814
FLORA Some additions to the Oklahoma flora. Rhodora 52: 19-24, 35-41. {} 815
FLORA Some results of a third summer's botanizing in Oklahoma. Rhodora 52: 816
FLORA More additions to the Oklahoma flora. Rhodora 53: 130. {} 817
FLORA Further studies of the Oklahoma flora. Rhodora 54: 125-131. {} 818
FLORA The identity of Hedyotis rosea Raf. Rhodora 55: 201-203. {NC flora} 819
FLORA Studies in the composition and distribution of the Oklahoma flora. XX 820
FLORA A new species of Carex (Section Phyllostachyae) from Oklahoma. Rhodor 821
FLORA Changes in status and new combinations for certain taxa in the Oklaho 822
FLORA Monarda stipitatoglandulosa, a new species from Oklahoma. Rhodora 72: 823
FLORA Keys to the flora of Oklahoma, 5th edition. Published by the author. 824

FLORA Some geographic relationships of the vascular flora of the Oklahoma O 825
FLORA A list of the vascular flora of the Oklahoma Ozarks. Proc. Okla. Acad 826
FLORA The geographic variation of Asclepias tuberosa interior in Oklahoma. 828
FLORA New and interesting plants from the Great Plains. Southw. Natur. 18: 834
FLORA A study of the flora of Oklahoma. M. S. Thesis, Univ. Okla., Norman. 840
FLORA Studies in the Danthonieae (Poaceae). Ph. D. Thesis, Univ. Okla., Nor 854

FLORISTIC A floristic study of the vascular plants of the Gypsum Hills and 35
FLORISTIC Floristic components of the Gypsum Hills and Redbed Plains of sou 36
FLORISTIC A floristic study of the woody vegetation of the North American C 302
FLORISTIC The floristic affinities of the Arbuckle Mountains in Oklahoma. A 328
FLORISTIC A study of the environment and floristic composition of an oak-hi 426
FLORISTIC A floristic study of the native or naturalized vegetation of Wash 459
FLORISTIC A floristic study of Washington county, Oklahoma. Proc. Okla. Aca 460

FLOWERS Notes on the phenology of middle western spring flowers. Proc. Okla 268
FLOWERS Phenological notes for 1928 on middle western spring flowers. Proc. 269
FLOWERS Oklahoma spring flowers in autumn. Proc. Okla. Acad. Sci. 17: 60-62 325
FLOWERS Roadside flowers of Oklahoma. C & J Printing Co., Lawton, OK. {flor 450
FLOWERS Roadside flowers of Oklahoma. Vol. 2. C & J Printing Co., Lawton, O 451

FORAGE Production characteristics of Oklahoma forages: Native range. Okla. 291
FORAGE Effects of chemical composition of soil on micronutrients and other 298
FORAGE Forage production, vegetative composition, and plant vigor in relati 304
FORAGE A summary of fire and forage research on shinnery oak rangelands. Pr 468

FORBS Nutrient analysis of selected forbs on clearcut areas in southeastern 594

FOREST Introgressive hybridization between shortleaf and loblolly pine in s 1
FOREST Some factors influencing the frequency of bark lichens in north cent 5
FOREST Some factors influencing the frequency of bark lichens in north cent 6
FOREST The effect of host specificity on the interspecific associations of 7
FOREST The effect of cement-kiln dust on vegetation. M. S. Thesis, Univ. Ok 21
FOREST Comparative water relations of four species of upland deciduous tree 23
FOREST Growth and distribution of Platanus occidentalis in Oklahoma as infl 30
FOREST The woody vegetation of Bear's Glen, a Washington Irving stopover. P 40
FOREST The influence of varying light intensities on the vegetative growth 49
FOREST Forest statistics for southeast Oklahoma counties – 1986. USDA Fore 51
FOREST Forest statistics for northeast Oklahoma – 1986. USDA Forest Ser. R 52
FOREST The short grass plains and post oak-blackjack woodland of Oklahoma i 56
FOREST Bottomland hardwoods of eastern Oklahoma: A special study of their s 72
FOREST Deciduous forests of eastern North America. Blakiston, Philadelphia. 74
FOREST Comparative studies of the animal communities of oak-hickory forests 76
FOREST The vegetation of Oklahoma. Ecol. Monogr. 1: 99-188. {forest tree gr 79
FOREST Relationships of the woody vegetation of the Wichita Mountains Wildl 84
FOREST Relationships of the woody vegetation of the Wichita Mountains Wildl 85
FOREST Some additions to Eskew's list of flowering plants of the Wichita Mo 90
FOREST Ecological studies in an eastern Oklahoma floodplain. Proc. Okla. Ac 93
FOREST The changing phytosociology of the oak-hickory forest of the Ozark M 117
FOREST Reproduction of woody plants in the absence of grazing. Proc. Okla. 119
FOREST Growth of ten species of preferred deer browse plants under various 132
FOREST The arborescent flora of Oklahoma. M. S. Thesis, Univ. Wyo., Laramie 134
FOREST Hydrologic and nutrient relationships in a pine-hardwood forest in s 137
FOREST Ordination and classification of mature bottomland forests in northc 140
FOREST An evaluation of the quadrat method in the blackjack-post oak forest 152

FOREST Effects of various plant and environmental factors on post oak respo 159
FOREST Low intensity forest sampling through use of stem frequency distribu 167
FOREST Final report on vegetation and microenvironments of Platt National P 171
FOREST The vegetation of the Wichita Mountains Wildlife Refuge. M. S. Thesi 186
FOREST Plant associations as site predictors in the pine-hardwood tension z 187
FOREST Genecology of eastern redbud (Cercis canadensis). Ecology 63: 962-97 190
FOREST Description and dynamics of some western oak forests in Oklahoma. Ph 191
FOREST Ordination and classification of western oak forests in Oklahoma. Am 192
FOREST Identifying the prairie-forest transition using remote sensing. M. S 196
FOREST A comparison of post oak-blackjack oak communities in two major soil 199
FOREST The savannah concept and its use. Ecology 38: 435-442. {classificati 200
FOREST Herbicides for release of shortleaf pine and native grass. Weeds 15: 207
FOREST Bio-economic analysis of even aged shortleaf pine stands in southeas 211
FOREST A quantitative comparison of the soil algae of central Oklahoma prai 213
FOREST The flowering plants of the Wichita National Forest. M. S. Thesis, U 216
FOREST Natural variation in rooting ability of western provenances of short 223
FOREST Silting and forest succession on Deep Fork in southwestern Creek Cou 229
FOREST Woodland of Indian Territory. U. S. Geol. Surv. Rep. No. 21: 603-672 235
FOREST The Cross Timbers. Star Printery, Muskogee, OK. {upland forest tree} 238
FOREST Silvics of forest trees of the United States. Agr. Handbook No. 271, 241
FOREST Petroleum and surface vegetation. Proc. Okla. Acad. Sci. 10: 110-114 267
FOREST Commerce of the prairies: or, The journal of a Santa Fe trader, duri 274
FOREST Economical harvesting systems for scattered stands of cottonwood in 278
FOREST Variation and hybridization in Juniperus. Ann. Mo. Bot. Gard. 39: 1- 281
FOREST Differential selection in juniper populations from the Baum Limeston 282
FOREST A comparative study of the water relations of Quercus marilandica an 283
FOREST Geographic distribution of two species of oaks in Oklahoma in relati 284
FOREST A preliminary survey of important factors which affect tree developm 292
FOREST Effect of silting on tree development in the flood plain of the Nort 293
FOREST Relation of climatic conditions, soil characteristics, and tree deve 294
FOREST Effect of silt on natural vegetation and drainage in the flood plain 297
FOREST A comparison of forest vegetation on orphan strip mines and adjacent 299
FOREST A floristic study of the woody vegetation of the North American Cros 302
FOREST Ecological studies on the Canadian River floodplain in Cleveland Cou 309
FOREST A study of interspecific hybridization in a natural oak population w 315
FOREST Interspecific hybridization in a natural oak population with particu 316
FOREST The effect of water potential upon the geographic distribution of Ca 318
FOREST Aspection in a central Oklahoma tension zone area. M. S. Thesis, Uni 320
FOREST Geology of the southern Ouachita Mountains of Oklahoma. Okla. Geol. 324
FOREST The effect of fire on the growth of Rhus glabra L. and Smilax bona-n 332
FOREST Vegetation in relation to slope exposure and geology in the Arbuckle 335
FOREST Some vegetation-environment relationships in the upland forests of O 344
FOREST Biomass, annual net primary production and dynamics of six mineral e 345
FOREST Vegetational changes in a black willow forest over a 23-year period. 346
FOREST Woody vegetation of southeastern LeFlore County, Oklahoma, in relati 347
FOREST Some vegetation-environment relationships in the upland forests of O 349
FOREST Correlation analysis of rainfall and annual ring index of central Ok 350
FOREST Biomass, annual net primary production, and dynamics of six mineral 351
FOREST A quantitative comparison between an oak forest and an oak savannah 352
FOREST Habitat preferences among representative wintering and breeding bird 353
FOREST An analysis of selected Oklahoma upland forest stands including both 375
FOREST Energy analyses of forestry practices in oak-pine forests and loblol 383
FOREST The effects of periodic flooding on upland forest succession. M. S. 387

FOREST The vegetation of the Caddo County Canyons, Oklahoma. Ecology 20: 1- 410
FOREST Acer grandidentatum in Oklahoma. Rhodora 46: 445-450. {SW Canadian C 411
FOREST Baldcypress (Taxodium distichum) in Oklahoma. Proc. Okla. Acad. Sci. 413
FOREST Forest trees of Oklahoma, 12th. edition. Okla. For. Div., State Dept 414
FOREST An ecological study of the Southeastern Oklahoma Protective Unit. Un 415
FOREST Trees and shrubs of the Southeastern Oklahoma Protective Unit. Proc. 416
FOREST Ecological factors affecting the distribution of woody vegetation ne 420
FOREST The detection and nature of functional underground connections betwe 422
FOREST Net production relations of two tree species in southern Oklahoma. P 424
FOREST A study of the environment and floristic composition of an oak-hicko 426
FOREST Relation between soil fungi and seed plants in three successional fo 437
FOREST Relation between soil fungi and seed plants in three successional fo 438
FOREST Efficiencies of the point-center quarter and quadrat methods in fore 446
FOREST Phenotypic variation and relationships of several variables in short 458
FOREST Competitive and allelopathic suppression of understory by Oklahoma o 476
FOREST Germination studies of "Caddo" sugar maple (Acer saccharum) seeds. S 482
FOREST Eleven year biologic-economic evaluation of undesirable hardwood sit 489
FOREST The Gulf Coastal Plain Ultisol and Alfisol ecosystem: surface geolog 502
FOREST Forest type conversion on the "Cross Timbers" area in Oklahoma. M. S 524
FOREST Relationship of soil properties to site index of shortleaf pine and 526
FOREST The ligneous flora of Rich Mountain, Arkansas and Oklahoma. J. Arnol 527
FOREST Edaphic and topographic effects on forest communities in Payne Count 536
FOREST An analysis of an elm-ash floodplain community near Norman, Oklahoma 541
FOREST The composition of a black willow forest in south central Oklahoma. 547
FOREST The composition of a post oak forest in south-central Oklahoma. Sout 549
FOREST Vegetational changes in a black willow forest over a four year perio 553
FOREST An evaluation of the arms-length rectangle method in forest sampling 559
FOREST A study of the role of Rhus copallina in the community dynamics of t 565
FOREST The role of Rhus copallina in the dynamics of the forest-prairie eco 566
FOREST A quantitative analysis of bottomland communities in south-central O 567
FOREST Nutrient analysis of selected forbs on clearcut areas in southeaster 594
FOREST The microclimate of a relict stand of sugar maple in Devil's Canyon 600
FOREST The microclimate of sugar maple stands in Oklahoma. Ecology 43: 19-2 601
FOREST Vegetation of Beaver's Bend State Park. Okla. Geol. Surv. Guide Book 602
FOREST Bottomland forests of north-central Oklahoma. Ecology 46: 708-714. { 605
FOREST Inhibition of nitrification by climax ecosystems. Amer. J. Bot. 59: 615
FOREST Inhibition of nitrification by climax ecosystems. II. Additional evi 616
FOREST Inhibition of nitrification by climax ecosystems. III. Inhibitors ot 617
FOREST An evaluation of the variable-radius and paired-tree methods in the 619
FOREST Composition of a green ash forest near Norman, Oklahoma. Southw. Nat 620
FOREST The upland forests of Oklahoma. Ecology 40: 593-608. {all forest} 621
FOREST Diversity in tree species in Oklahoma upland forests. Ecology 52: 87 632
FOREST Phytosociological analysis of Oklahoma upland forest species. Ecolog 633
FOREST Sediment-precipitation-runoff relationships for the three ephemeral 638
FOREST A short phytogeography of the prairies and Great Plains of central N 646
FOREST Factor analysis of tree distribution patterns in Oklahoma. Ecology 5 656
FOREST Shrub distribution patterns in Oklahoma with comparisons to those ex 657
FOREST Deer browse in the Ouachita forest in Oklahoma. J. Wildl. Manage. 32 658
FOREST Dispersion and co-dispersion in Oklahoma upland forests. Bull. Torre 678
FOREST Oklahoma forests and woodlands. Proc. Okla. Acad. Sci. 30: 3-10. {al 684
FOREST Some variations of the fruit of Quercus macrocarpa. M. S. Thesis, Ok 708
FOREST Cottonwood site evaluation in Oklahoma. M. S. Thesis, Okla. State Un 715
FOREST The relation of forest vegetation to soils and geology in the gulf c 729

FOREST Stem frequency distribution and sampling of the hardwood component o 760
FOREST Descriptive and hydrologic characteristics of three forested watersh 801
FOREST The comparative ecology of two canyons and an upland area in west ce 837
FOREST Journal of the expedition of the regiment of dragoons into Indian co 839
FOREST Atlas of the woody plants of Oklahoma. Okla. Biol. Surv., Norman. {t 845
FOREST A survey of the soil algae of two vegetation types in central Oklaho 847
FOREST Shortleaf pine site index – Soils and plant associations on the coa 851
FOREST Environmental parameters and site identification for potential expan 853
FOREST Natural areas for Oklahoma: important areas of natural significance 860

FOREST-PRAIRIE Habitat preferences among representative wintering and breed 353
FOREST-PRAIRIE A study of the role of Rhus copallina in the community dynam 565
FOREST-PRAIRIE The role of Rhus copallina in the dynamics of the forest-pra 566

FORMATION Relationships of the woody vegetation of the Wichita Mountains Wi 84
FORMATION Relationships of the woody vegetation of the Wichita Mountains Wi 85
FORMATION The relation of stock pond vegetation to geological formations in 145
FORMATION Systematic study of the flowering plants in the vicinity of Norma 392
FORMATION Correlation between vegetation and geologic formations in Oklahom 597

FRAXINUS An analysis of an elm-ash floodplain community near Norman, Oklaho 541

FREQUENCY Some factors influencing the frequency of bark lichens in north c 5
FREQUENCY Some factors influencing the frequency of bark lichens in north c 6
FREQUENCY Low intensity forest sampling through use of stem frequency distr 167
FREQUENCY Stem frequency distribution and sampling of the hardwood componen 760

FRESHWATER Electrophoretic analysis of soluble enzymes in five freshwater d 303
FRESHWATER A survey of auxotrophy in five freshwater dinoflagellates (Pyrrh 322

FRUIT Roadside wild fruits of Oklahoma. Univ. Okla. Press, Norman. {flora} 452
FRUIT Some variations of the fruit of Quercus macrocarpa. M. S. Thesis, Okl 708
FRUIT Notes on the growth and fruiting of Andropogon scoparius, with some r 806

FUNGI A comparison of the soil fungi of a tall-grass prairie and of an aban 212
FUNGI A comparison of soil microfungi of two grassland areas in central Okl 288
FUNGI Relation between soil fungi and seed plants in three successional for 437
FUNGI Relation between soil fungi and seed plants in three successional for 438
FUNGI A comparison of the soil fungi of a virgin prairie and an abandoned f 461
FUNGI Fungi of various depths in typical Cleveland County, Oklahoma soils. 763

GALIUM Galium pedemontanum (Bell.) All. (Rubiaceae) reported from Oklahoma. 395
GALIUM The genus Galium (Rubiaceae) in the southeastern United States. Ph. 396

GALLIC Inhibition of symbiotic nitrogen-fixation by gallic and gallotannic 57
GALLIC Inhibition of symbiotic nitrogen fixation by gallic and tannic acid, 58

GALLOTANNIC Inhibition of symbiotic nitrogen-fixation by gallic and gallota 57

GARDENING The ecological basis of ornamental gardening in Oklahoma. M. S. T 501

GATHERING Gathering of Ipomopsis rubra (Polemoniaceae) nectar by Bombus (Ap 775

GENECOLOGY Genecology of eastern redbud (Cercis canadensis). Ecology 63: 96 190

GENERIC The generic concept and circumscription in the Triticeae: an end pa 221

GENETIC Introgressive hybridization between shortleaf and loblolly pine in 1
GENETIC The genetic relationship between Nama hispidum and Nama stevensii. 215
GENETIC Genetic variation in juvenile characteristics of Populus deltoides 497

GEOGRAPHIC A comparative study of the water relations of Quercus marilandic 283
GEOGRAPHIC Geographic distribution of two species of oaks in Oklahoma in re 284
GEOGRAPHIC The effect of water potential upon the geographic distribution o 318
GEOGRAPHIC A biogeographical boundary: The Tatschl Line. Trans. Kans. Acad. 389
GEOGRAPHIC The Chisholm Trail in geographic perspective. Proc. Okla. Acad. 761
GEOGRAPHIC Some geographic relationships of the vascular flora of the Oklah 825
GEOGRAPHIC The geographic variation of Asclepias tuberosa interior in Oklah 828
GEOGRAPHIC Biogeographic regions of Texas and Oklahoma. Ecology 31: 427-433 833

GEOLOGIC Relationships of the woody vegetation of the Wichita Mountains Wil 84
GEOLOGIC Relationships of the woody vegetation of the Wichita Mountains Wil 85
GEOLOGIC The relation of stock pond vegetation to geological formations in 145
GEOLOGIC Correlation between vegetation and geologic formations in Oklahoma 597
GEOLOGIC Geological map of Oklahoma. USGS, Washington. {geology all} 785

GEOLOGY Geology and petrology of the Wichita Mountains. Okla. Geol. Surv. B 319
GEOLOGY Geology of the southern Ouachita Mountains of Oklahoma. Okla. Geol. 324
GEOLOGY Vegetation in relation to slope exposure and geology in the Arbuckl 335
GEOLOGY The Gulf Coastal Plain Ultisol and Alfisol ecosystem: surface geolo 502
GEOLOGY The Arbuckle Mountains, Oklahoma. Okla. Geol. Surv. Circ. No. 14. { 595
GEOLOGY Correlation between vegetation and geologic formations in Oklahoma. 597
GEOLOGY Distribution of legumes as correlated with surface geology and plan 673
GEOLOGY The relation of forest vegetation to soils and geology in the gulf 729
GEOLOGY Geological map of Oklahoma. USGS, Washington. {geology all} 785

GERMINATION After-ripening and germination of seed of Juniperus virginiana 160
GERMINATION Germination studies of "Caddo" sugar maple (Acer saccharum) see 482
GERMINATION Germination of downy brome from southern Kansas, central Oklaho 483
GERMINATION Germination of weed seeds with reference to temperatures and st 493
GERMINATION Mechanism of seed dormancy in Ambrosia artemisiifolia. Ph. D. D 844

GLEDITSIA Honeylocust (Gleditsia triacanthos L.) in shelterbelts of western 243

GLOEODENDRON Gloeodendron: A rare alga in Oklahoma. Proc. Okla. Acad. Sci. 855

GLOTTIDIUM Glottidium vesicarium in Oklahoma. Rhodora 36: 118-119. {flora} 260

GLUCOSE Acetylene reduction (N-fixation) of glucose amended soils from cent 361

GOLD Marcy and the gold seekers. Univ. Okla. Press, Norman. {presettlement} 240

GOLDENROD New species, new combinations, and notes on the goldenrods (Eutha 725
GOLDENROD Solidago flaccidifolia Small (Asteraceae) new to Oklahoma and oth 798

GRADIENT Comparative water relations of four species of upland deciduous tr 23
GRADIENT An analysis of the vegetation of buffalo wallows in relation to en 579

GRAMA A new species of grama grass. Bot. Gaz. 91: 103-105. {Comanche SW flo 225
GRAMA Prairie dogs, whitefaces, and blue grama. Wildl. Monogr. 3: 1-78. {gr 386

GRAMINEAE A study of the angiosperms, exclusive of the Gramineae, Cyperacea 153
GRAMINEAE Distribution of the grasses (Gramineae) in relation to plate tect 520
GRAMINEAE A systematic and ecological study of the Gramineae, Cyperaceae, a 588
GRAMINEAE The spermatophyta of Cherokee County, Oklahoma (exclusive of 802

GRASS Plant inhibition by Johnson grass and its possible significance in pl 2
GRASS Plant inhibition by Johnson grass and its possible significance in ol 3
GRASS Nutrient and biomass allocation strategies for five grass species in 8
GRASS Differential response of woody and herbaceous species to summer and w 9

GRASS Pollination systems in Paspalum dilatatum Poir. (Poaceae): An example 10
GRASS Nutrient and biomass allocation in five grass species in an Oklahoma 11
GRASS Tall grass defeats prairie dogs. Soil Cons. 20:103-105. {SW grassland 22
GRASS The short grass plains and post oak-blackjack woodland of Oklahoma in 56
GRASS Analysis of morphological variation in Schizachyrium scoparium (littl 77
GRASS A study of the native grasses of Comanche County, Oklahoma. M. S. The 130
GRASS Some morphological and physiological effects of density and density-w 197
GRASS Water relations of some grasses with phreatophytic properties. Ph. D. 205
GRASS New methods of brush control for more grass. J. Range Manage. 3: 46-5 210
GRASS Grasses and grasslands: systematics and ecology. Univ. Okla. Press, N 218
GRASS The grasses and grasslands of Oklahoma. Status of current research on 219
GRASS Classification of Oklahoma grasses. Ann. Okla. Acad. Sci. 6: 141-161. 220
GRASS The generic concept and circumscription in the Triticeae: an end pape 221
GRASS A new species of grama grass. Bot. Gaz. 91: 103-105. {Comanche SW flo 225
GRASS Grasses of Oklahoma. Okla. A & M Coll. Tech. Bull. No. 3. {flora all} 226
GRASS The identification of some of the more common native Oklahoma grasses 301
GRASS Developmental growth of four species of range grasses in north-centra 334
GRASS Prairie dogs, whitefaces, and blue grama. Wildl. Monogr. 3: 1-78. {gr 386
GRASS Grass stand establishment on rough, rocky, wooded lands in eastern Ok 423
GRASS The seeding, development, and utilization of native grasses in Kay Co 485
GRASS Hybridization and introgression between Elymus canadensis and Elymus 498
GRASS The physiological ecology of upland and lowland Panicum virgatum. Ph. 505
GRASS Distribution of the grasses (Gramineae) in relation to plate tectonic 520
GRASS Replacement of a population of Johnson grass by a vine-forb community 556
GRASS A comparative study of upland and bottomland types of Panicum virgatu 582
GRASS An analysis of variation between upland and lowland switchgrass, Pani 583
GRASS Growth and development of five species of range grasses in central Ok 598
GRASS Competitive responses of Bouteloua curtipendula (Michx.) Torr. Amer. 624
GRASS Carbon dioxide exchange characteristics of some prairie grass seedlin 630
GRASS Methods of reestablishing buffalo grass on cultivated land in the Gre 647
GRASS Drought survival of native grass species in the central and southern 648
GRASS Possible mechanisms of the exclusion of Johnson grass by tall grass p 660
GRASS Some effects of differential clipping on six native grasses and on on 692
GRASS Allelopathic inhibition of weed invasion by Andropogon scoparius Mich 713
GRASS Notes on the growth and fruiting of Andropogon scoparius, with some r 806

GRASSLAND Plant inhibition by Johnson grass and its possible significance i 2
GRASSLAND Plant inhibition by Johnson grass and its possible significance i 3
GRASSLAND Nutrient and biomass allocation strategies for five grass species 8
GRASSLAND Differential response of woody and herbaceous species to summer a 9
GRASSLAND Nutrient and biomass allocation in five grass species in an Oklah 11
GRASSLAND The response of a central Oklahoma grassland to burning. Southw. 12
GRASSLAND Phenology of a native tall-grass prairie in central Oklahoma. Eco 13
GRASSLAND Phenology of a native tall-grass prairie in central Oklahoma. M. 14
GRASSLAND Tall grass defeats prairie dogs. Soil Cons. 20:103-105. {SW grass 22
GRASSLAND Distribution and control of several woody plants in Texas and Okl 25
GRASSLAND Vegetation-soil relationships in Flint Hills bluestem pastures. J 27
GRASSLAND Flowering patterns and production in a central Oklahoma grassland 28
GRASSLAND Pasture and range plants. Phillips Petrol. Co., Bartlesville, OK. 29
GRASSLAND Conservation ranching in the Oklahoma panhandle. J. Range Manage. 31
GRASSLAND A preliminary study of buffalo wallows in the vicinity of Norman, 45
GRASSLAND Evaluation of sampling techniques on tall-grass prairie. J. Range 48
GRASSLAND Ecological relationships of the Bird Creek region, northeastern O 54

GRASSLAND The short grass plains and post oak-blackjack woodland of Oklahom 56
GRASSLAND Standing state and cycling of nitrogen in soil-vegetation compone 61
GRASSLAND The climate of the central North American Grassland. Ann. Assoc. 68
GRASSLAND Analysis of morphological variation in Schizachyrium scoparium (l 77
GRASSLAND Overgrazing from an ecological point of view. Proc. Okla. Acad. S 78
GRASSLAND The vegetation of Oklahoma. Ecol. Monogr. 1: 99-188. {forest tree 79
GRASSLAND A partial inventory of insect populations in tallgrass prairie pa 81
GRASSLAND Microclimates of three grassland plots in central Oklahoma. M. S. 82
GRASSLAND A survey of the tall grass prairies in northeastern Oklahoma. M. 83
GRASSLAND Some additions to Eskew's list of flowering plants of the Wichita 90
GRASSLAND A survey of the tall-grass prairie in northeastern Oklahoma. Sout 91
GRASSLAND Micronutrients in soils of the tall-grass prairie of Oklahoma. Pr 120
GRASSLAND Seasonal aspects in certain prairie and savannah communities. M. 121
GRASSLAND Some observations on the number and size of samples necessary to 122
GRASSLAND Plant distribution in a central Oklahoma prairie. Proc. Okla. Aca 123
GRASSLAND Fluctuations in biotic communities. III. Aspection in a ravine se 124
GRASSLAND Fluctuations in biotic communities. V. Aspection in mixed-grass p 125
GRASSLAND Ecology of the grassland biome. M. S. Thesis, Univ. Okla., Norman 126
GRASSLAND The grassland biome. Ecol. Monogr. 10: 617-684. {prairie shortgra 127
GRASSLAND A study of the native grasses of Comanche County, Oklahoma. M. S. 130
GRASSLAND Vegetation analysis of bobwhite habitat of northeastern Oklahoma 138
GRASSLAND Interaction of disturbances in tallgrass prairie: a field experim 139
GRASSLAND Succession in grasslands: thirty-two years of change in a central 141
GRASSLAND The effect of early spring burning on vegetation in buffalo wallo 142
GRASSLAND Seed predation, seed dispersal, and disturbance in grasslands: a 143
GRASSLAND Effects of disturbance on diversity in mixed-grass prairie. Veget 144
GRASSLAND Vegetation structure in a tallgrass prairie. M. S. Thesis, Univ. 146
GRASSLAND Canopy structure of a tall-grass prairie. J. Range Manage. 27: 31 147
GRASSLAND Insect distribution and seasonal succession in overgrazed and nor 157
GRASSLAND Preliminary studies in the evaluation of grassland sampling techn 161
GRASSLAND Grassland communities of the Wichita Mountains Wildlife Refuge. P 162
GRASSLAND Influence of soils and parent materials on grasslands of the Wich 163
GRASSLAND Effect of mowing on a relict tall grass prairie. Proc. Okla. Acad 164
GRASSLAND The grasslands of Platt National Park, Oklahoma. Southw. Natur. 4 170
GRASSLAND Final report on vegetation and microenvironments of Platt Nationa 171
GRASSLAND The black-tailed prairie dog (Cynomys ludovicianus) as a member o 176
GRASSLAND The vegetation of the Wichita Mountains Wildlife Refuge. M. S. Th 186
GRASSLAND Some morphological and physiological effects of density and densi 197
GRASSLAND An annotated plant list for Adams Ranch. M. S. Thesis, Ft. Hays K 198
GRASSLAND Influence of grazing on plant succession of rangelands. Bot. Rev. 204
GRASSLAND New methods of brush control for more grass. J. Range Manage. 3: 210
GRASSLAND Grasses and grasslands: systematics and ecology. Univ. Okla. Pres 218
GRASSLAND The grasses and grasslands of Oklahoma. Status of current researc 219
GRASSLAND Classification of Oklahoma grasses. Ann. Okla. Acad. Sci. 6: 141- 220
GRASSLAND The generic concept and circumscription in the Triticeae: an end 221
GRASSLAND Interseeding and paraquat effects on central and eastern rangelan 224
GRASSLAND Carrying capacity of pastures in Muskogee County, Oklahoma. M. S. 231
GRASSLAND Rodent populations and their reactions to grazing intensities on 242
GRASSLAND Effects of mowing on a native tall grass prairie in central Oklah 247
GRASSLAND A vegetational analysis of the Phillips Agricultural Demonstratio 248
GRASSLAND Why prairies are treeless. Science N. S. 22: 55-56. {fire burning 249
GRASSLAND The effects of nitrogen fertilization on native vegetation under 250
GRASSLAND The effects of one year's nitrogen fertilization on native vegeta 251

GRASSLAND Burning and fertilization for range improvement in central Oklaho 271
GRASSLAND A comparison of soil microfungi of two grassland areas in central 288
GRASSLAND Production characteristics of Oklahoma forages: Native range. Okl 291
GRASSLAND Effects of chemical composition of soil on micronutrients and oth 298
GRASSLAND The identification of some of the more common native Oklahoma gra 301
GRASSLAND Forage production, vegetative composition, and plant vigor in rel 304
GRASSLAND The claypan range site in northern Osage County, Oklahoma. J. Ran 305
GRASSLAND Effect of grazing intensity on plant composition, vigor, and prod 306
GRASSLAND Retrogression of native range vegetation in the sandy high plains 317
GRASSLAND Effect of fertilization on native grass pastures in Oklahoma. J. 331
GRASSLAND Developmental growth of four species of range grasses in north-ce 334
GRASSLAND Vegetation in relation to slope exposure and geology in the Arbuc 335
GRASSLAND Mechanisms of secondary succession in a tall grass prairie. Ph. D 336
GRASSLAND Changes in species composition of grassland communities in respon 337
GRASSLAND A tour of the prairies. In crayon miscellany. Carey, Lea, and Bla 338
GRASSLAND Acetylene reduction (N-fixation) in soil and old-field succession 362
GRASSLAND Symbiotic and asymbiotic N2-fixation in a tall grass prairie. Soi 363
GRASSLAND Vegetation and soil conditions of prairie and pasture plots in ce 367
GRASSLAND Effects of moderate grazing on the composition and plant producti 369
GRASSLAND Winter burning in central Oklahoma grassland. Ecology 38: 520-522 370
GRASSLAND Carthamus lanatus L. (Asteraceae: Cynareae) a potentially serious 376
GRASSLAND Restoration of eroded prairie with digested sludge. Proc. Okla. A 377
GRASSLAND Grazing use checks on the Wichitas. J. Range Manage. 16: 142-145. 378
GRASSLAND Energy analysis of Oklahoma rangelands and improved pastures. J. 384
GRASSLAND Prairie dogs, whitefaces, and blue grama. Wildl. Monogr. 3: 1-78. 386
GRASSLAND A biogeographical boundary: The Tatschl Line. Trans. Kans. Acad. 389
GRASSLAND Nitrogen fixation in selected prairie legumes as related to succe 419
GRASSLAND Grass stand establishment on rough, rocky, wooded lands in easter 423
GRASSLAND Microbial decomposition of cellulose and native plant litter in a 447
GRASSLAND Populations and range effects of rodents on the sand sagebrush gr 457
GRASSLAND A comparison of the soil fungi of a virgin prairie and an abandon 461
GRASSLAND An investigation of the soil microflora of two grassland plots. P 462
GRASSLAND Spraying 2,4-D by airplane on sand sagebrush and other plants of 469
GRASSLAND Fourteen year summary of range improvement studies. Bull. U. S. S 470
GRASSLAND The role of ecotypic variation in the distribution of the central 473
GRASSLAND Range and pasture studies and grazing recommendations for the sou 474
GRASSLAND Data and observations concerning regrassing abandoned cultivated 475
GRASSLAND The seeding, development, and utilization of native grasses in Ka 485
GRASSLAND A phenological comparison of mowed and unmowed tallgrass prairie 488
GRASSLAND A vegetation analysis of a pimpled prairie in northeastern Oklaho 491
GRASSLAND Range deterioration in south central Oklahoma. M. S. Thesis, Univ 494
GRASSLAND Possible role of Ambrosia psilostachya on pattern and succession 496
GRASSLAND The effect of disturbances in tall grass prairie sites on an inde 503
GRASSLAND The physiological ecology of upland and lowland Panicum virgatum. 505
GRASSLAND Revegetation of eroded lands in the High Plains of Texas and Okla 508
GRASSLAND Relative effects of known plant inhibitors on species from first 514
GRASSLAND Some experiments and observations on revegetation of Oklahoma ran 515
GRASSLAND Vegetation of an abandoned prairie dog town in tall-grass prairie 523
GRASSLAND The effect of grazing on habitat utilization of the dickcissel (S 525
GRASSLAND Ecological study of western Payne County, Oklahoma land utilizati 535
GRASSLAND A comparison of the insects of a virgin prairie, a grazed pasture 537
GRASSLAND Simulated impact of management practices upon the tallgrass prair 540
GRASSLAND A modification of the point-centered quarter method for grassland 550

GRASSLAND The relation of grazing to plant succession in the tall grass pra 551
GRASSLAND Effects of denudation on the productivity of grassland. Ecology 4 552
GRASSLAND Replacement of a population of Johnson grass by a vine-forb commu 556
GRASSLAND Plant population changes in a native prairie plot plowed annually 558
GRASSLAND Some effects of winter burning on a moderately grazed pasture. Ec 560
GRASSLAND Some aspects of structure and function in Oklahoma old-field succ 562
GRASSLAND Some aspects of structure and function in Oklahoma old-field succ 563
GRASSLAND The distribution of rodents in overgrazed and normal grasslands i 575
GRASSLAND Rodent distribution in overgrazed and normal grasslands. M. S. Th 576
GRASSLAND The relationship of plant species heterogeneity to soil variation 580
GRASSLAND Relationships of vegetation and environment in buffalo wallows. A 581
GRASSLAND An analysis of variation between upland and lowland switchgrass, 583
GRASSLAND The vegetation regions of the prairie province. Bot. Gaz. 25: 381 585
GRASSLAND Oak (Quercus spp.) sprout growth rates on a central Oklahoma shal 586
GRASSLAND Establishment and management of tame and native grass pastures in 587
GRASSLAND A phytosociological analysis of the tall-grass prairie in northea 592
GRASSLAND A phytosociological analysis of the tall-grass prairie in northea 593
GRASSLAND Phytosociological analysis of a tall-grass prairie in Marshall Co 599
GRASSLAND Inhibition of nodulation of inoculated legumes by pioneer plant s 607
GRASSLAND Allelopathy and grassland improvement. In: Estes, J. R., and R. J 608
GRASSLAND Seed dispersal and mineral nutrition in succession in abandoned f 611
GRASSLAND Causes of decreases in residual carbohydrase activity in soil dur 612
GRASSLAND Causes of decreases in productivity in undisturbed tall grass pra 613
GRASSLAND Inhibition of nitrification by climax ecosystems. Amer. J. Bot. 5 615
GRASSLAND Inhibition of nitrification by climax ecosystems. II. Additional 616
GRASSLAND Inhibition of nitrification by climax ecosystems. III. Inhibitors 617
GRASSLAND Plant succession and yield of living plant material in a plowed p 618
GRASSLAND Competitive responses of Bouteloua curtipendula (Michx.) Torr. Am 624
GRASSLAND Preliminary producer data synthesis: 1970 comprehensive network s 625
GRASSLAND Osage site, 1970 report: Primary production. USIBP Grassland Biom 626
GRASSLAND Plant community structure. In: N. R. French (ed) Preliminary anal 627
GRASSLAND A preliminary compartment model of a tall grass prairie: Osage 19 628
GRASSLAND Systems analysis of a tall grass prairie. Proc. Third Midw. Prair 629
GRASSLAND Herbage dynamics of a tall grass prairie: Osage, 1971. USIBP Gras 634
GRASSLAND Herbage dynamics of an Oklahoma tallgrass prairie: Osage 1972. US 635
GRASSLAND Ecosystem analysis of the tallgrass prairie: nitrogen cycle. Ecol 636
GRASSLAND The true prairie ecosystem. Hutchinson Ross Publ. Co., Stroudsbur 637
GRASSLAND Plant succession on abandoned fields in central Oklahoma and in t 643
GRASSLAND A short phytogeography of the prairies and Great Plains of centra 646
GRASSLAND Methods of reestablishing buffalo grass on cultivated land in the 647
GRASSLAND Drought survival of native grass species in the central and south 648
GRASSLAND The Southern Great Plains. Grass: USDA Yearbook of Agric., p. 503 649
GRASSLAND Natural revegetation of abandoned farmland in the central and sou 650
GRASSLAND Observations on the grasslands of the central United States. Ohio 651
GRASSLAND Ecology of the scaled quail in the Oklahoma panhandle. Wildl. Mon 653
GRASSLAND Possible mechanisms of the exclusion of Johnson grass by tall gra 660
GRASSLAND Establishment of warm and cool season grass pastures on wooded si 664
GRASSLAND The natural vegetation of the Great Plains region. Ann. Assoc. Am 668
GRASSLAND Calcium, magnesium, phosphorus, and potassium budgets of a tall g 674
GRASSLAND A model for phosphorus and potassium flux in a tall-grass prairie 675
GRASSLAND The pattern of retrogression of native vegetation in north centra 686
GRASSLAND The structure and function of ten western North American grasslan 687
GRASSLAND Pattern of retrogression of native vegetation in north central Ok 688

GRASSLAND The structure and function of ten western North American grasslan 689
GRASSLAND The structure and function of ten North American grasslands. III. 690
GRASSLAND The structure and function of ten western North American grasslan 691
GRASSLAND Some effects of differential clipping on six native grasses and o 692
GRASSLAND Effect of disturbance on microclimate and vegetation in tall gras 693
GRASSLAND The effect of overgrazing and erosion upon the biota of the mixed 696
GRASSLAND Biotic and physiographic succession on abandoned eroded farmland. 697
GRASSLAND Allelopathic inhibition of weed invasion by Andropogon scoparius 713
GRASSLAND The influence of burning on tillering and aboveground biomass par 717
GRASSLAND A continuation of the investigation of the soil microflora of two 718
GRASSLAND A broad analysis of fifteen sites in the tall-grass prairie of Ok 719
GRASSLAND Some observations on the spread of mesquite to the north in Cimar 720
GRASSLAND The grassland communities of the Baum Limestone in Johnston Count 755
GRASSLAND The Chisholm Trail in geographic perspective. Proc. Okla. Acad. S 761
GRASSLAND Notes on the growth and fruiting of Andropogon scoparius, with so 806
GRASSLAND Progress of regrassing in the Southern Great Plains. Soil Cons. 5 829
GRASSLAND North American prairie. Johnsen Pub. Co., Lincoln, Neb. {grasslan 830
GRASSLAND Grasslands of the Great Plains. Johnsen Pub. Co., Lincoln, Neb. { 831
GRASSLAND Effects of excessive mulch on development, yield, and structure o 832
GRASSLAND The effect of overgrazing on insect populations. Proc. Okla. Acad 836
GRASSLAND A reconnaisance of a typical central Oklahoma prairie. M. S. Thes 838
GRASSLAND A survey of the soil algae of two vegetation types in central Okl 847
GRASSLAND An exploratory study on soil algae. Ecology 38: 309-313. {SC gras 848
GRASSLAND Natural areas for Oklahoma: important areas of natural significan 860

GRAY'S Gray's manual of botany, 8th edition. American Book Co., New York. { 233

GRAZED Insect distribution and seasonal succession in overgrazed and normal 157
GRAZED A comparison of the insects of a virgin prairie, a grazed pasture, a 537
GRAZED Some effects of winter burning on a moderately grazed pasture. Ecolo 560
GRAZED The distribution of rodents in overgrazed and normal grasslands in c 575
GRAZED Rodent distribution in overgrazed and normal grasslands. M. S. Thesi 576

GRAZING Overgrazing from an ecological point of view. Proc. Okla. Acad. Sci 78
GRAZING Reproduction of woody plants in the absence of grazing. Proc. Okla. 119
GRAZING Interaction of disturbances in tallgrass prairie: a field experimen 139
GRAZING Insect distribution and seasonal succession in overgrazed and norma 157
GRAZING Influence of grazing on plant succession of rangelands. Bot. Rev. 2 204
GRAZING Rodent populations and their reactions to grazing intensities on sa 242
GRAZING The effects of nitrogen fertilization on native vegetation under cl 250
GRAZING Effect of grazing intensity on plant composition, vigor, and produc 306
GRAZING Changes in species composition of grassland communities in response 337
GRAZING Effects of moderate grazing on the composition and plant production 369
GRAZING Grazing use checks on the Wichitas. J. Range Manage. 16: 142-145. { 378
GRAZING Range and pasture studies and grazing recommendations for the south 474
GRAZING The effect of grazing on habitat utilization of the dickcissel (Spi 525
GRAZING The relation of grazing to plant succession in the tall grass prair 551
GRAZING Rodent distribution in overgrazed and normal grasslands. M. S. Thes 576
GRAZING The effect of overgrazing and erosion upon the biota of the mixed g 696
GRAZING The effect of overgrazing on insect populations. Proc. Okla. Acad. 836

GROUNDSELS Toxic plants of Oklahoma: Groundsels and rattlebox. J. Okla. Vet 104

GROWTH Growth and distribution of Platanus occidentalis in Oklahoma as infl 30
GROWTH The influence of varying light intensities on the vegetative growth 49

GROWTH Growth of ten species of preferred deer browse plants under various 132
GROWTH Vegetation response following winged elm (Ulmus alata Michx.) and oa 172
GROWTH The effect of fire on the growth of Rhus glabra L. and Smilax bona-n 332
GROWTH Developmental growth of four species of range grasses in north-centr 334
GROWTH Effects of bacterial inhibitors and certain related compounds on the 513
GROWTH Effects of certain plants of old-field succession on the growth of b 538
GROWTH Oak (Quercus spp.) sprout growth rates on a central Oklahoma shallow 586
GROWTH Growth and development of five species of range grasses in central O 598
GROWTH Notes on the growth and fruiting of Andropogon scoparius, with some 806

GULF The Gulf Coastal Plain Ultisol and Alfisol ecosystem: surface geology, 502
GULF Distribution, seasonal abundance, and hosts of the Gulf Coast tick in 662
GULF The relation of forest vegetation to soils and geology in the gulf coa 729

GULLY-CONTROL Revegetation and gully-control experiments in the Red Plain 209

GYPSUM A floristic study of the vascular plants of the Gypsum Hills and Red 35
GYPSUM Floristic components of the Gypsum Hills and Redbed Plains of southw 36

HABITAT The comparative utility of plant life-form and other veg. chars. in 50
HABITAT Bottomland hardwoods of eastern Oklahoma: A special study of their 72
HABITAT Comparative studies of the animal communities of oak-hickory forest 76
HABITAT Life history and habitat requirements of burrowing owls in western 115
HABITAT Nesting of the mourning dove in northwest Oklahoma. Proc. Okla. Aca 128
HABITAT Ecology of the nesting birds of the McCurtain Game Preserve, Oklaho 129
HABITAT Growth of ten species of preferred deer browse plants under various 132
HABITAT The life-form concept and mule deer habitat in Oklahoma. Proc. Okla 135
HABITAT Vegetation analysis of bobwhite habitat of northeastern Oklahoma ra 138
HABITAT Insect distribution and seasonal succession in overgrazed and norma 157
HABITAT Components of the habitat of the bobwhite quail in Payne County, Ok 175
HABITAT The black-tailed prairie dog (Cynomys ludovicianus) as a member of 176
HABITAT Evaluation of Payne County, Oklahoma farmlands and vegetation patte 177
HABITAT Manipulating central Oklahoma rangeland for bobwhite quail. M. S. T 185
HABITAT Brush control and the welfare of lesser prairie chickens in western 188
HABITAT Effects on lesser prairie chickens of brush control in western Okla 189
HABITAT An evaluation of ground nesting by mourning doves in northwestern O 193
HABITAT A survey of the game and furbearing animals of Oklahoma. Okla. Game 195
HABITAT Rodent populations and their reactions to grazing intensities on sa 242
HABITAT The influence of habitat quality on density of cotton rat populatio 258
HABITAT Plants for improving bobwhite habitat in northwestern Oklahoma. Okl 289
HABITAT A preliminary report on the seasonal aspects of six habitats near N 308
HABITAT Habitat preferences among representative wintering and breeding bir 353
HABITAT The life-form concept applied to prairie chicken habitat in Oklahom 354
HABITAT Identification and analysis of lesser and greater prairie chi 356
HABITAT Habitat used by lesser prairie chickens for feeding related to seas 357
HABITAT Prairie dogs, whitefaces, and blue grama. Wildl. Monogr. 3: 1-78. { 386
HABITAT The effect of land use on nest sites selected by Mississippi kites. 421
HABITAT The mammals of Harmon County, Oklahoma. Proc. Okla. Acad. Sci. 49: 444
HABITAT Populations and range effects of rodents on the sand sagebrush gras 457
HABITAT Preliminary observations on mourning doves wintering in southwester 487
HABITAT The effect of grazing on habitat utilization of the dickcissel (Spi 525
HABITAT A comparison of the insects of a virgin prairie, a grazed pasture, 537
HABITAT The distribution of rodents in overgrazed and normal grasslands in 575
HABITAT Rodent distribution in overgrazed and normal grasslands. M. S. Thes 576
HABITAT Deer browse in the Ouachita forest in Oklahoma. J. Wildl. Manage. 3 658

HABITAT The ecology and behavior of the lone star tick (Acarina: Ixodidae). 661
HABITAT Distribution, seasonal abundance, and hosts of the Gulf Coast tick 662
HABITAT The ecology and behavior of the lone star tick. (Acarina: Ixodidae) 663
HABITAT Distribution, habitat, and reproductive biology of Phlox oklahomens 703
HABITAT Habitat description and the life-form concept. Proc. Okla. Acad. Sc 707
HABITAT Habitat types used by the bobwhite quail and other wildlife in sout 772
HABITAT Vertebrates in a prairie dog town. Proc. Okla. Acad. Sci. 50: 110-1 773

HARDWOOD Bottomland hardwoods of eastern Oklahoma: A special study of their 72
HARDWOOD Hydrologic and nutrient relationships in a pine-hardwood forest in 137
HARDWOOD Plant associations as site predictors in the pine-hardwood tension 187
HARDWOOD Eleven year biologic-economic evaluation of undesirable hardwood s 489
HARDWOOD Stem frequency distribution and sampling of the hardwood component 760

HARVESTING Economical harvesting systems for scattered stands of cottonwood 278

HAY Seed hay for regrassing eroded land. Proc. Okla. Acad. Sci. 25: 43-45. 206
HAY Hay fever plants in Oklahoma. Ann. Allerg. 45: 26-32. {allergy} 399
HAY Non-allergenic native and cultivated plants in Oklahoma: Landscaping wi 400
HAY Effects of oat straw and native hay mulches on plant composition and ed 610

HEDYOTIS Hedyotis ouachitana (Rubiaceae): a new species from the Ouachita M 698
HEDYOTIS The identity of Hedyotis rosea Raf. Rhodora 55: 201-203. {NC flora 819

HELIANTHUS A study of allelopathy in Helianthus rigidus. M. S. Thesis, Okla 790
HELIANTHUS Allelopathy as expressed by Helianthus annuus, and its role in o 849
HELIANTHUS Allelopathy as expressed by Helianthus annuus and its role in ol 850

HEMLOCK Toxic plants of Oklahoma: Hemlocks. J. Okla. Vet. Med. Assoc. 34: 3 100

HERBACEOUS Differential response of woody and herbaceous species to summer 9
HERBACEOUS An outline for ecological life histories of herbaceous vascular 542
HERBACEOUS Herbaceous plants. In: Rare and endangeied vertebrates and plant 700

HERBAGE Herbage dynamics of a tall grass prairie: Osage, 1971. USIBP Grassl 634
HERBAGE Herbage dynamics of an Oklahoma tallgrass prairie: Osage 1972. USIB 635

HERBARIUM Notes from the herbarium of the University of Oklahoma. Rhodora 4 327
HERBARIUM Notes from the Bebb Herbarium of the University of Oklahoma II. R 329

HERBICIDES Herbicides for release of shortleaf pine and native grass. Weeds 207

HERITAGE The Oklahoma Natural Heritage Program 1977-1984: a critical evalua 480
HERITAGE Natural areas for Oklahoma: important areas of natural significanc 860

HIGHVELD Plant succession on abandoned fields in central Oklahoma and in th 643

HISTORICAL The short grass plains and post oak-blackjack woodland of Oklaho 56

HISTORIES An outline for ecological life histories of herbaceous vascular h 542

HISTORY Life history and habitat requirements of burrowing owls in western 115
HISTORY The Oklahoma botanical travels of G. W. Stevens. Proc. Okla. Acad. 264
HISTORY The botanical itineraries of A. H. Van Vleet. Chron. Okla. 56: 322- 265
HISTORY Early botanists of Oklahoma. M. S. Thesis, Okla. A & M Coll., Still 310
HISTORY Adaptation of the life-history of Schinia mitis to its host plant P 862

HONEY Honeylocust (Gleditsia triacanthos L.) in shelterbelts of western Okl 243
HONEY The ecology of Honey Creek: Temporal patterns of the travertine perip 596

HONEYLOCUST Honeylocust (Gleditsia triacanthos L.) in shelterbelts of weste 243

HORSETAILS Toxic plants of Oklahoma: Ferns and horsetails. J. Okla. Vet. Me 106

HOUSEPLANTS Toxic plants of Oklahoma: Ornamental houseplants. J. Okla. Vet. 99

HYBRIDIZATION Introgressive hybridization between shortleaf and loblolly pi 1
HYBRIDIZATION Variation and hybridization in Juniperus. Ann. Mo. Bot. Gard. 281
HYBRIDIZATION A study of interspecific hybridization in a natural oak popul 315
HYBRIDIZATION Interspecific hybridization in a natural oak population with 316
HYBRIDIZATION Hybridization and introgression between Elymus canadensis and 498

HYDROGEN A comparison of organic carbon, hydrogen ion concentration, and vo 368
HYDROGEN Soil enzymes in relation to old field succession: amylase, cellula 531

HYDROLOGIC Hydrologic and nutrient relationships in a pine-hardwood forest 137
HYDROLOGIC Descriptive and hydrologic characteristics of three forested wat 801

HYDROLOGY Sediment-precipitation-runoff relationships for the three ephemer 638

HYDROPHYLLACEAE A biosystematic study of the relationship of Nama hispidum 779

HYDROPHYTE An outline for ecological life histories of herbaceous vascular 542

HYPOXIS Hypoxis longii. Status report. Endangered Species Office, U. S. Fis 744
HYPOXIS Biology and distribution of Hypoxis longii Fern. (Amaryllidaceae). 753

IDENTIFICATION Distribution and identification of woody plants of Oklahoma 86
IDENTIFICATION The identification of some of the more common native Oklahom 301
IDENTIFICATION Identification and analysis of lesser and greater prairie ch 356
IDENTIFICATION Inhibition of nitrogen-fixing and nitrifying bacteria by see 604
IDENTIFICATION Identification and mapping of the extant flora at the Deer C 776
IDENTIFICATION Environmental parameters and site identification for potenti 853

IDENTITY The identity of Hedyotis rosea Raf. Rhodora 55: 201-203. {NC flora 819

IMPACT Simulated impact of management practices upon the tallgrass prairie. 540
IMPACT Endangered and threatened plants of Oklahoma. Dept. Botany-Micro., U 859

IMPORTANCE Algae as pioneers in plant succession and their importance in er 66
IMPORTANCE The increasing importance of vines in southern Oklahoma. Proc. O 539

IMPORTANT A preliminary survey of important factors which affect tree devel 292
IMPORTANT Natural areas for Oklahoma: important areas of natural significan 860

IMPROVED Energy analysis of Oklahoma rangelands and improved pastures. J. R 384

IMPROVEMENT A soil improvement program for Oklahoma. Okla. A & M Coll. C 131
IMPROVEMENT Burning and fertilization for range improvement in central Okla 271
IMPROVEMENT Fourteen year summary of range improvement studies. Bull. U. S. 470
IMPROVEMENT Allelopathy and grassland improvement. In: Estes, J. R., and R. 608

INDIAN A list of some of the most interesting species of plants collected i 114
INDIAN Woodland of Indian Territory. U. S. Geol. Surv. Rep. No. 21: 603-672 235
INDIAN List of plants collected by C. S. Sheldon and M. A. Carlton in the I 323
INDIAN Journal of the expedition of the regiment of dragoons into Indian co 839

INDICATOR Natural vegetation as an indicator of the capabilities of land fo 667
INDICATOR Epiphytic diatoms as indicators of water quality in farm and expe 765
INDICATOR Epiphytic diatoms as indicators of water quality in farm and expe 768
INDICATOR Life zones and zone indicators in Oklahoma. Proc. Okla. Acad. Sci 841

INFLUENCE Growth and distribution of Platanus occidentalis in Oklahoma as i 30

INFLUENCE The influence of varying light intensities on the vegetative grow 49
INFLUENCE Influence of soils and parent materials on grasslands of the Wich 163
INFLUENCE Influence of grazing on plant succession of rangelands. Bot. Rev. 204
INFLUENCE The influence of habitat quality on density of cotton rat populat 258
INFLUENCE Influence of a wildfire in the Wichita Mountains Wildlife Refuge. 554
INFLUENCE The influence of burning on tillering and aboveground biomass par 717

INHIBITION Plant inhibition by Johnson grass and its possible significance 2
INHIBITION Plant inhibition by Johnson grass and its possible significance 3
INHIBITION Ecological studies of plant inhibition by Solanum rostratum. Ph. 15
INHIBITION Inhibition of symbiotic nitrogen-fixation by gallic and gallotan 57
INHIBITION Inhibition of symbiotic nitrogen fixation by gallic and tannic a 58
INHIBITION The possible ecological significance of inhibition of Euphorbia 75
INHIBITION Inhibitional effects of Digitaria sanguinalis and possible role 533
INHIBITION Inhibitional effects of Digitaria sanguinalis and possible role 534
INHIBITION Inhibition of nitrogen-fixing and nitrifying bacteria by seed pl 603
INHIBITION Inhibition of nitrogen-fixing and nitrifying bacteria by seed pl 604
INHIBITION Inhibition of nodulation of inoculated legumes by pioneer plant 607
INHIBITION Inhibition of nitrification by climax ecosystems. Amer. J. Bot. 615
INHIBITION Inhibition of nitrification by climax ecosystems. II. Additional 616
INHIBITION Inhibition of nitrification by climax ecosystems. III. Inhibitor 617
INHIBITION Allelopathic inhibition of weed invasion by Andropogon scoparius 713

INHIBITORS Effects of bacterial inhibitors and certain related compounds on 513
INHIBITORS Relative effects of known plant inhibitors on species from first 514
INHIBITORS Inhibition of nitrogen-fixing and nitrifying bacteria by seed pl 604
INHIBITORS Inhibition of nitrification by climax ecosystems. III. Inhibitor 617

INJURY Injury and death or recovery of trees in a prairie climate. Ecol. Mo 20

INTERSEEDING Interseeding and paraquat effects on central and eastern range 224

INTERSPECIFIC The effect of host specificity on the interspecific associati 7
INTERSPECIFIC A study of interspecific hybridization in a natural oak popul 315
INTERSPECIFIC Interspecific hybridization in a natural oak population with 316

INTRA-SEASONAL The structure and function of ten western North American gra 691

INTRODUCED The grasses and grasslands of Oklahoma. Status of current resear 219
INTRODUCED Some effects of differential clipping on six native grasses and 692

INTROGRESSION A study of interspecific hybridization in a natural oak popul 315
INTROGRESSION Interspecific hybridization in a natural oak population with 316
INTROGRESSION Hybridization and introgression between Elymus canadensis and 498

INTROGRESSIVE Introgressive hybridization between shortleaf and loblolly pi 1

INVASION Allelopathic inhibition of weed invasion by Andropogon scoparius M 713

INVENTORY A partial inventory of insect populations in tallgrass prairie pa 81
INVENTORY A biological inventory of the Fort Gibson Lake area. U. S. Dept. 796

IPOMOPSIS Gathering of Ipomopsis rubra (Polemoniaceae) nectar by Bombus (Ap 775

IRRIGATION An environmental study of the W. C. Austin Irrigation Project of 314

IRVING The woody vegetation of Bear's Glen, a Washington Irving stopover. P 40
IRVING On the western tour with Washington Irving: The journal and letters 701
IRVING Washington Irving on the prairie, or, a narrative of a tour in the s 846

ISOLATION Isolation of plants in the Oklahoma flora. Proc. Okla. Acad. Sci. 326
ISOLATION Loss of ferulic acid from decomposing leaves of Celtis laevigata 770

ITINERARIES The botanical itineraries of A. H. Van Vleet. Chron. Okla. 56: 265

JUNCACEAE A systematic and ecological study of the Gramineae, Cyperaceae, a 588

JUNIPER Differential selection in juniper populations from the Baum Limesto 282
JUNIPER The vegetation of the Mesa de Maya region of Colorado, New Mexico, 639
JUNIPER Some botanical studies in the Black Mesa region of Oklahoma. Rhodor 640

JUNIPERUS After-ripening and germination of seed of Juniperus virginiana L. 160
JUNIPERUS A taxonomic study of the genus Juniperus in Oklahoma. M. S. Thesi 280
JUNIPERUS Variation and hybridization in Juniperus. Ann. Mo. Bot. Gard. 39: 281
JUNIPERUS Wind patterns and the seasonal occurrence of Juniperus pollen in 402

KAW Identification and mapping of the extant flora at the Deer Creek archae 776
KAW Final updated and revised environmental statement: Kaw Lake, Arkansas R 784

KENTUCKY Toxic plants of Oklahoma: Oak, black locust, Kentucky coffee-tree, 96

KEYS Keys to the flora of Oklahoma, 5th edition. Published by the author. { 824

KIAMICHI Vascular plants of southeastern Oklahoma from the Sans Bois to the 478
KIAMICHI Final environmental statement: Hugo Lake, Kiamichi River, Oklahoma 783
KIAMICHI Descriptive and hydrologic characteristics of three forested water 801

LABIATAE A numerical taxonomic study of Monarda (Labiatae) in southern Okla 441

LAKE Water quality and phytoplankton communities in Lake Thunderbird, Cleve 47
LAKE Some algae of Lake Carl Blackwell, Oklahoma, exclusive of diatoms. M. 149
LAKE Water fluctuation as a factor in the life of the higher plants of a 33 181
LAKE The drowning of trees in Lake Carl Blackwell. Proc. Okla. Acad. Sci. 2 228
LAKE Observations on the effects of erosion on the ecology and maintenance 358
LAKE The algae of Crystal Lake, Cleveland County, Oklahoma. Amer. Midl. Nat 397
LAKE The algae of Crystal Lake, 1938-1978. M. S. Thesis, Univ. Okla., Norma 427
LAKE Crystal Lake revisited. Proc. Okla. Acad. Sci. 61: 49-57. {SC Clevelan 428
LAKE A limnological study of two artificial lakes in Oklahoma. M. S. Thesis 516
LAKE Plant communities of Oklahoma lakes. Ecology 34: 561-583. {aquatic} 543
LAKE A limnological study of two small artificial lakes. M. S. Thesis, Univ 762
LAKE Identification and mapping of the extant flora at the Deer Creek archa 776
LAKE Final environmental statement: Optima Lake, North Canadian River, Okla 782
LAKE Final environmental statement: Hugo Lake, Kiamichi River, Oklahoma. Tu 783
LAKE Final updated and revised environmental statement: Kaw Lake, Arkansas 784
LAKE A biological inventory of the Fort Gibson Lake area. U. S. Dept. of th 796

LANDSCAPES Ecological interpretation for small mounds in landscapes of east 24

LEAD Plant communities and the potential for ecotypic differentiation on le 254
LEAD The natural revegetation of lead/zinc mine spoil in northeastern Oklah 255

LEAD/ZINC The natural revegetation of lead/zinc mine spoil in northeastern 255

LEAVES Leaves from a collector's notebook. J. Arnold Arb. 13:417-437. {flor 529
LEAVES Juvenile leaves in Oklahoma Marsilea (Marsileaceae). Sida 7: 218-219 759
LEAVES Loss of ferulic acid from decomposing leaves of Celtis laevigata and 770

LEGUMES Nitrogen fixation in selected prairie legumes as related to success 419
LEGUMES Inhibition of nodulation of inoculated legumes by pioneer plant spe 607
LEGUMES Distribution of legumes as correlated with surface geology and plan 673

LEGUMINOSEAE Allium ampeloprasum (Liliaceae) and Trifolium vesiculosum (Leg 757

LEMNAEA Lemnaea fuscina Borg (Rhodophyta) in Oklahoma. Proc. Okla. Acad. Sc 730

LESQUERELLA Lesquerella angustifolia. Status report. Endangered Species Off 748

LICHENS Some factors influencing the frequency of bark lichens in north cen 5
LICHENS Some factors influencing the frequency of bark lichens in north cen 6
LICHENS The effect of host specificity on the interspecific associations of 7
LICHENS Lichens from the state of Oklahoma. Pap. Mich. Acad. Sci. Arts. Let 307
LICHENS The lichens of north central Oklahoma. Ph. D. Diss., Okla. State Un 365

LIFE-FORM The comparative utility of plant life-form and other veg. chars. 50
LIFE-FORM The life-form concept and mule deer habitat in Oklahoma. Proc. Ok 135
LIFE-FORM The life-form concept applied to prairie chicken habitat in Oklah 354
LIFE-FORM Habitat description and the life-form concept. Proc. Okla. Acad. 707

LIGHT The influence of varying light intensities on the vegetative growth o 49

LIGNEOUS The ligneous flora of Rich Mountain, Arkansas and Oklahoma. J. Arn 527

LILIACEAE Lilium michiganense Farwell (Liliaceae) and Gentiana alba Muhl. (46
LILIACEAE Allium ampeloprasum (Liliaceae) and Trifolium vesiculosum (Legumi 757

LIMESTONE Differential selection in juniper populations from the Baum Limes 282
LIMESTONE The grassland communities of the Baum Limestone in Johnston Count 755

LIMNOLOGICAL Plankton production and species distribution in the limnologic 379
LIMNOLOGICAL A limnological study of two artificial lakes in Oklahoma. M. S 516
LIMNOLOGICAL A limnological study of two small artificial lakes. M. S. Thes 762

LISTERA Listera australis Lindley (Orchidaceae) new for Oklahoma. Southw. N 431

LITERATURE Literature on the vegetation of Oklahoma. Proc. Okla. Acad. Sci. 372
LITERATURE Literature on the vegetation of Oklahoma, 1964-1975. Proc. Okla. 481
LITERATURE Additions to the literature on the vegetation of Oklahoma. Proc. 484
LITERATURE Abstracts of ecological literature pertaining to Oklahoma. M. S. 622

LITTER Microbial decomposition of cellulose and native plant litter in a tr 447

LIVERWORT The liverworts of Oklahoma. Bryologist 39: 25-34. {flora all} 407
LIVERWORT Bryophytes of eastern Oklahoma. Proc. Okla. Acad. Sci. 9: 35-38. 670
LIVERWORT The bryophytes of Oklahoma: A preliminary study with notes on the 671
LIVERWORT Bryophyta of southeast Oklahoma. I. A preliminary list with notes 672

LOBLOLLY Introgressive hybridization between shortleaf and loblolly pine in 1
LOBLOLLY Energy analyses of forestry practices in oak-pine forests and lobl 383
LOBLOLLY Natural regeneration of loblolly pine 230 miles northwest of its n 584

LOCOWEED Toxic plants of Oklahoma: Locoweeds. J. Okla. Vet. Med. Assoc. 33 95

LOUISIANA Exploration of the Red River of Louisiana. Senate Ex. Doc. No. 54 443

LOWLAND The physiological ecology of upland and lowland Panicum virgatum. P 505
LOWLAND An analysis of variation between upland and lowland switchgrass, Pa 583

MACROPHYTES Aquatic plants of Oklahoma: I. Submersed, floating-leaved, and 499

MALVACEAE Reproductive studies in the genus Callirhoe (Malvaceae). Ph. D. T 26

MANAGEMENT Distribution and control of several woody plants in Texas and Ok 25
MANAGEMENT Conservation ranching in the Oklahoma panhandle. J. Range Man 31

MANAGEMENT The tallgrass prairie orchids Platanthera leucophaea (Nutt.) Lin 70
MANAGEMENT A soil improvement program for Oklahoma. Okla. A & M Coll. Ci 131
MANAGEMENT Effects of various plant and environmental factors on post oak r 159
MANAGEMENT Effect of mowing on a relict tall grass prairie. Proc. Okla. Aca 164
MANAGEMENT Vegetation response following winged elm (Ulmus alata Michx.) an 172
MANAGEMENT Vegetational responses following winged elm and oak control in O 173
MANAGEMENT Investigations in erosion control and reclamation of eroded land 174
MANAGEMENT A preliminary study of the larger aquatic plants of Oklahoma wit 179
MANAGEMENT A preliminary study of the larger aquatic plants of Oklahoma wit 180
MANAGEMENT Manipulating central Oklahoma rangeland for bobwhite quail. M. S 185
MANAGEMENT Plant associations as site predictors in the pine-hardwood tensi 187
MANAGEMENT Brush control and the welfare of lesser prairie chickens in west 188
MANAGEMENT Effects on lesser prairie chickens of brush control in western O 189
MANAGEMENT Seed hay for regrassing eroded land. Proc. Okla. Acad. Sci. 25: 206
MANAGEMENT Herbicides for release of shortleaf pine and native grass. Weeds 207
MANAGEMENT Brush control and pasture development in the Red Plains. Agron. 208
MANAGEMENT Revegetation and gully-control experiments in the Red Plains Re 209
MANAGEMENT New methods of brush control for more grass. J. Range Manage. 3: 210
MANAGEMENT Bio-economic analysis of even aged shortleaf pine stands in sout 211
MANAGEMENT Interseeding and paraquat effects on central and eastern rangela 224
MANAGEMENT Carrying capacity of pastures in Muskogee County, Oklahoma. M. S 231
MANAGEMENT The mesquite problem in the southwest. J. Range Manage. 3: 60-71 234
MANAGEMENT Rodent populations and their reactions to grazing intensities on 242
MANAGEMENT Honeylocust (Gleditsia triacanthos L.) in shelterbelts of wester 243
MANAGEMENT Effects of mowing on a native tall grass prairie in central Okla 247
MANAGEMENT The effects of nitrogen fertilization on native vegetation under 250
MANAGEMENT The effects of one year's nitrogen fertilization on native veget 251
MANAGEMENT Burning and fertilization for range improvement in central Oklah 271
MANAGEMENT Economical harvesting systems for scattered stands of cottonwood 278
MANAGEMENT Plants for improving bobwhite habitat in northwestern Oklahoma. 289
MANAGEMENT Production characteristics of Oklahoma forages: Native range. Ok 291
MANAGEMENT A limited-range provenance test of open-pollenated progeny of Ca 300
MANAGEMENT Forage production, vegetative composition, and plant vigor in re 304
MANAGEMENT Effect of grazing intensity on plant composition, vigor, and pro 306
MANAGEMENT Retrogression of native range vegetation in the sandy high plain 317
MANAGEMENT Effect of fertilization on native grass pastures in Oklahoma. J. 331
MANAGEMENT Changes in species composition of grassland communities in respo 337
MANAGEMENT Effects of moderate grazing on the composition and plant product 369
MANAGEMENT Winter burning in central Oklahoma grassland. Ecology 38: 520-52 370
MANAGEMENT Restoration of eroded prairie with digested sludge. Proc. Okla. 377
MANAGEMENT Energy analyses of forestry practices in oak-pine forests and lo 383
MANAGEMENT The effect of land use on nest sites selected by Mississippi kit 421
MANAGEMENT Grass stand establishment on rough, rocky, wooded lands in easte 423
MANAGEMENT Shinnery oak can be controlled. Proc. Southern Weed Conf. 9: 95- 466
MANAGEMENT Progress in shinnery oak and sand sage control at Woodward. Prog 467
MANAGEMENT A summary of fire and forage research on shinnery oak rangelands 468
MANAGEMENT Spraying 2,4-D by airplane on sand sagebrush and other plants of 469
MANAGEMENT Fourteen year summary of range improvement studies. Bull. U. S. 470
MANAGEMENT Field surveys for noxious weeds in Oklahoma, Project 12-16-5-230 477
MANAGEMENT Eleven year biologic-economic evaluation of undesirable hardwood 489
MANAGEMENT Some experiments and observations on revegetation of Oklahoma 515
MANAGEMENT Simulated impact of management practices upon the tallgrass prai 540
MANAGEMENT Effects of denudation on the productivity of grassland. Ecology 552

MANAGEMENT Effects of fencing and plowing on plant succession in a revegeta 557
MANAGEMENT Plant population changes in a native prairie plot plowed annuall 558
MANAGEMENT Establishment and management of tame and native grass pastures i 587
MANAGEMENT Nutrient analysis of selected forbs on clearcut areas in southea 594
MANAGEMENT Allelopathy and grassland improvement. In: Estes, J. R., and R. 608
MANAGEMENT Methods of reestablishing buffalo grass on cultivated land in th 647
MANAGEMENT Establishment of warm and cool season grass pastures on wooded s 664
MANAGEMENT Some effects of differential clipping on six native grasses and 692
MANAGEMENT Cottonwood site evaluation in Oklahoma. M. S. Thesis, Okla. Stat 715
MANAGEMENT Shortleaf pine site index – Soils and plant associations on the 851
MANAGEMENT Environmental parameters and site identification for potential e 853

MAP A game type map of Oklahoma. Okla. Game and Fish Dept., Okla. City. {cl 194
MAP Silvics of forest trees of the United States. Agr. Handbook No. 271, US 241
MAP Potential natural vegetation of the conterminous United States. Amer. G 388
MAP Biotic type mapping of Oklahoma watersheds. Proc. Okla. Acad. Sci. 22: 521
MAP Identification and mapping of the extant flora at the Deer Creek archae 776
MAP Geological map of Oklahoma. USGS, Washington. {geology all} 785
MAP Atlas of the woody plants of Oklahoma. Okla. Biol. Surv., Norman. {tree 845

MAPLE Relationships of two isolated groups of sugar maple in central Oklaho 183
MAPLE Relationships of two isolated groups of sugar maples (Acer saccharum 184
MAPLE Relationships of sugar maples (Acer saccharum and A. grandidentatum) 252
MAPLE Germination studies of "Caddo" sugar maple (Acer saccharum) seeds. Se 482
MAPLE The microclimate of a relict stand of sugar maple in Devil's Canyon i 600
MAPLE The microclimate of sugar maple stands in Oklahoma. Ecology 43: 19-25 601

MAPPING Biotic type mapping of Oklahoma watersheds. Proc. Okla. Acad. Sci. 521
MAPPING Identification and mapping of the extant flora at the Deer Creek ar 776

MARCY Adventures on the Red River. A journal of Capt. R. B. Marcy. Univ. Ok 239
MARCY Marcy and the gold seekers. Univ. Okla. Press, Norman. {presettlement 240
MARCY Report of Capt. R. B. Marcy on the railroad from Fort Smith to Santa 442

MARINE A third species of Enteromorpha (a marine Chlorophycean) for Oklahom 148
MARINE Enteromorpha, a marine alga in Oklahoma. Proc. Okla. Acad. Sci. 56: 569

MARSH Salinity tolerance of marsh plants in Kansas and Oklahoma. Ecology 47 791

MARSILEA Juvenile leaves in Oklahoma Marsilea (Marsileaceae). Sida 7: 218-2 759

MARTYNIACEAE The reproductive biology of Proboscidea louisianica (Martyniac 573
MARTYNIACEAE Floral biology of Proboscidea louisianica (Martyniaceae). Rhod 758

MATERIAL Influence of soils and parent materials on grasslands of the Wichi 163
MATERIAL Plant succession and yield of living plant material in a plowed pr 618

MATERIALS Influence of soils and parent materials on grasslands of the Wich 163

MATURE Ordination and classification of mature bottomland forests in northe 140

MAYA The vegetation of the Mesa de Maya region of Colorado, New Mexico, and 639

MCGEE Vegetation resources including endangered and threatened species of t 312
MCGEE Field investigations, McGee Creek Project, Oklahoma, Calamovilfa arcu 313

MEADOW A preliminary report on the ecology of a Drosera meadow. Proc. Okla. 39

MECHANISM Mechanisms of secondary succession in a tall grass prairie. Ph. D 336
MECHANISM Possible mechanisms of the exclusion of Johnson grass by tall gra 660

MECHANISM Mechanism of seed dormancy in Ambrosia artemisiifolia. Ph. D. Dis 844

MELANTHIUM Veratrum woodii (Melanthium woodii). Status report. Endangered S 742

MESA The vegetation of the Mesa de Maya region of Colorado, New Mexico, and 639
MESA Some botanical studies in the Black Mesa region of Oklahoma. Rhodora 5 640

MESQUITE The mesquite problem in the southwest. J. Range Manage. 3: 60-71. 234
MESQUITE The mammals of Harmon County, Oklahoma. Proc. Okla. Acad. Sci. 49: 444
MESQUITE A comparison of the predominant soil microflora in eight vegetativ 706
MESQUITE Some observations on the spread of mesquite to the north in Cimarr 720

METABOLIC The effect of moisture on the experimental determination of the m 158

METALLIFEROUS Evidence for the absence of ecotypic development in Andropogo 257

MICROBIAL Microbial decomposition of cellulose and native plant litter in a 447
MICROBIAL Condensed tannins in soil: inputs and effects on microbial popula 486
MICROBIAL Microbial decomposition of ferulic acid in soil. J. Chem. Ecol. 1 771

MICROCLIMATE Microclimates of three grassland plots in central Oklahoma. M. 82
MICROCLIMATE The microclimate of a relict stand of sugar maple in Devil's C 600
MICROCLIMATE The microclimate of sugar maple stands in Oklahoma. Ecology 43 601
MICROCLIMATE Effect of disturbance on microclimate and vegetation in tall g 693

MICROFLORA An investigation of the soil microflora of two grassland plots. 462
MICROFLORA A comparison of the predominant soil microflora in eight vegetat 706
MICROFLORA A continuation of the investigation of the soil microflora of tw 718

MICROFUNGI A comparison of soil microfungi of two grassland areas in centra 288

MICRONUTRIENTS Micronutrients in soils of the tall-grass prairie of Oklahom 120
MICRONUTRIENTS Effects of chemical composition of soil on micronutrients an 298

MICROORGANISMS The effect of moisture on the experimental determination of 158

MILFOIL Eurasian water milfoil in Oklahoma. Proc. Okla. Acad. Sci. 58: 4-5. 156
MILFOIL Eurasian watermilfoil – A new menace to Oklahoma waters. Proc. Okl 341

MILKWEEDS Toxic plants of Oklahoma: Milkweeds. J. Okla. Vet. Med. Assoc. 36 108

MIMA Ecological interpretation for small mounds in landscapes of eastern Ok 24

MINE Plant communities and the potential for ecotypic differentiation on le 254
MINE The natural revegetation of lead/zinc mine spoil in northeastern Oklah 255
MINE Revegetation of unreclaimed coal strip mines in Oklahoma. II. Plant co 256
MINE Evidence for the absence of ecotypic development in Andropogon virgini 257
MINE A comparison of forest vegetation on orphan strip mines and adjacent s 299
MINE Revegetation of unreclaimed coal strip-mines in Oklahoma. I. Vegetatio 348
MINE Soil factors preventing revegetation of a denuded area near an abandon 532
MINE Flora and fauna of Oklahoma abandoned mine lands. Okla. Biol. Surv., N 654
MINE Model fish and wildlife plan for a surface coal mine in Oklahoma. Okla 655
MINE Revegetation of strip mine spoils. M. S. Thesis, Okla. State Univ., St 702

MISSISSIPPI The effect of land use on nest sites selected by Mississippi ki 421

MIXED-GRASS Fluctuations in biotic communities. V. Aspection in mixed-grass 125
MIXED-GRASS Effects of disturbance on diversity in mixed-grass prairie. Veg 144

MODEL A preliminary compartment model of a tall grass prairie: Osage 1970. 628
MODEL Model fish and wildlife plan for a surface coal mine in Oklahoma. Okl 655
MODEL A model for phosphorus and potassium flux in a tall-grass prairie. So 675

MOISTURE Comparative water relations of four species of upland deciduous tr 23
MOISTURE The effect of moisture on the experimental determination of the me 158

MONARDA A numerical taxonomic study of Monarda (Labiatae) in southern Oklah 441
MONARDA Monarda stipitatoglandulosa. Status report. Endangered Species Offi 746
MONARDA Monarda stipitatoglandulosa, a new species from Oklahoma. Rhodora 7 823

MONOGRAPH A monographic study of the genus Aster in Oklahoma. M. S. Thesis, 236

MORPHOLOGICAL Analysis of morphological variation in Schizachyrium scopariu 77
MORPHOLOGICAL Cytological and morphological variation in Panicum virgatum L 80
MORPHOLOGICAL Some morphological and physiological effects of density and d 197
MORPHOLOGICAL A numerical analysis of the morphological variation in the ge 623

MOSS Mosses of Tulsa and Osage Counties. Proc. Okla. Acad. Sci. 8: 38-41. { 504
MOSS Bryophytes of eastern Oklahoma. Proc. Okla. Acad. Sci. 9: 35-38. {NE S 670
MOSS The bryophytes of Oklahoma: A preliminary study with notes on their ec 671
MOSS Bryophyta of southeast Oklahoma. I. A preliminary list with notes. Bry 672

MOUNDS Ecological interpretation for small mounds in landscapes of eastern 24

MOUNTAINS Relationships of the woody vegetation of the Wichita Mountains Wi 84
MOUNTAINS Relationships of the woody vegetation of the Wichita Mountains Wi 85
MOUNTAINS Some additions to Eskew's list of flowering plants of the Wichita 90
MOUNTAINS The changing phytosociology of the oak-hickory forest of the Ozar 117
MOUNTAINS Grassland communities of the Wichita Mountains Wildlife Refuge. P 162
MOUNTAINS Influence of soils and parent materials on grasslands of the Wich 163
MOUNTAINS A preliminary survey of the flora of the Arbuckle Mountains. M. S 168
MOUNTAINS A preliminary survey of the flora of the Arbuckle Mountains. Texa 169
MOUNTAINS The vegetation of the Wichita Mountains Wildlife Refuge. M. S. Th 186
MOUNTAINS The flowering plants of the Wichita Mountains Wildlife Refuge. Am 217
MOUNTAINS A preliminary list of the plants of the Arbuckle Mountains. B. A. 244
MOUNTAINS The occurrence of persimmon and buckeye in the Wichita Mountains 285
MOUNTAINS A bibliography of the Wichita Mountains of Oklahoma. Great Plains 286
MOUNTAINS Geology and petrology of the Wichita Mountains. Okla. Geol. Surv. 319
MOUNTAINS Geology of the southern Ouachita Mountains of Oklahoma. Okla. Geo 324
MOUNTAINS The floristic affinities of the Arbuckle Mountains in Oklahoma. A 328
MOUNTAINS Vegetation in relation to slope exposure and geology in the Arbuc 335
MOUNTAINS Account of an expedition from Pittsburg to the Rocky Mountains, p 340
MOUNTAINS Vascular plants of southeastern Oklahoma from the Sans Bois to th 478
MOUNTAINS A preliminary report on the flora of the Wichita Mountains. M. S. 492
MOUNTAINS Influence of a wildfire in the Wichita Mountains Wildlife Refuge. 554
MOUNTAINS The Arbuckle Mountains, Oklahoma. Okla. Geol. Surv. Circ. No. 14. 595
MOUNTAINS Hedyotis ouachitana (Rubiaceae): a new species from the Ouachita 698
MOUNTAINS Some account of a collection of plants made during a journey to a 764
MOUNTAINS Descriptive and hydrologic characteristics of three forested wate 801

MOWED A phenological comparison of mowed and unmowed tallgrass prairie site 488

MOWING Effect of mowing on a relict tall grass prairie. Proc. Okla. Acad. S 164
MOWING Effects of mowing on a native tall grass prairie in central Oklahoma 247

MULCH Effects of oat straw and native hay mulches on plant composition and 610
MULCH Effects of excessive mulch on development, yield, and structure of na 832

MYRIOPHYLLUM Myriophyllum in Oklahoma. Proc. Okla. Acad. Sci. 63: 103-104. 155

N-FIXATION Inhibition of symbiotic nitrogen-fixation by gallic and gallotan 57

N-FIXATION Acetylene reduction (N-fixation) of glucose amended soils from c 361
N-FIXATION Acctylene reduction (N-fixation) in soil and old-field successio 362

NAMA The genetic relationship between Nama hispidum and Nama stevensii. M. 215
NAMA A biosystematic study of the relationship of Nama hispidum and Nama st 276
NAMA A biosystematic study of the relationship of Nama hispidum and Nama st 779
NAMA Chromosome numbers in Nama and Proboscidea; IOPB Chromosome R 780

NATIONAL Platt National Park: Environment and ecology. Univ. Okla. Press, N 42
NATIONAL The effect of salinity on the distribution of algae at the Salt Pl 113
NATIONAL The grasslands of Platt National Park, Oklahoma. Southw. Natur. 4: 170
NATIONAL Final report on vegetation and microenvironments of Platt National 171
NATIONAL The flowering plants of the Wichita National Forest. M. S. Thesis, 216
NATIONAL National Wildlife Refuges - What future? Ecosystems 6: 4-11. {} 727
NATIONAL Nomenclature of the genus Achillea. In: National list of scientifi 778

NATURALIST Naturalist's guide to the Americas. Williams and Wilkins Co., Ba 676

NATURALIZED A study of the angiosperms, exclusive of the Gramineae, Cyperac 153
NATURALIZED A taxonomic study of the vascular plants indigenous to, or natu 455
NATURALIZED A floristic study of the native or naturalized vegetation of Wa 459

NATURE The Nature Conservancy's first Oklahoma preserve. Proc. Okla. Acad. 41
NATURE A nature study outline on trees and shrubs for the first six grades. 290
NATURE The detection and nature of functional underground connections betwe 422

NECTAR Gathering of Ipomopsis rubra (Polemoniaceae) nectar by Bombus (Apida 775

NESTS A study of viable seeds in a selection of birds' nests. Ph. D. Thesis 507

NIGHTSHADE Silverleaf nightshade (Solanum eleagnifolium) origin, distributi 71
NIGHTSHADE Toxic plants of Oklahoma: Thornapples and nightshades. J. Okla. 97

NITRATE Differences in nitrate reductase activity between species of differ 699

NITRIFICATION Some roles of allelopathic compounds in plant communities. Bi 609
NITRIFICATION Inhibition of nitrification by climax ecosystems. Amer. J. Bo 615
NITRIFICATION Inhibition of nitrification by climax ecosystems. II. Additio 616
NITRIFICATION Inhibition of nitrification by climax ecosystems. III. Inhibi 617

NITRIFYING Inhibition of nitrogen-fixing and nitrifying bacteria by seed pl 603
NITRIFYING Inhibition of nitrogen-fixing and nitrifying bacteria by seed pl 604

NITROGEN Inhibition of symbiotic nitrogen-fixation by gallic and gallotanni 57
NITROGEN Inhibition of symbiotic nitrogen fixation by gallic and tannic aci 58
NITROGEN Standing state and cycling of nitrogen in soil-vegetation componen 61
NITROGEN The effects of nitrogen fertilization on native vegetation under c 250
NITROGEN The effects of one year's nitrogen fertilization on native vegetat 251
NITROGEN Acetylene reduction (nitrogen fixation) and old-field succession i 359
NITROGEN Nitrogen fixation in selected prairie legumes as related to succes 419
NITROGEN Inhibition of nitrogen-fixing and nitrifying bacteria by seed plan 603
NITROGEN Inhibition of nitrogen-fixing and nitrifying bacteria by seed plan 604
NITROGEN Ecosystem analysis of the tallgrass prairie: nitrogen cycle. Ecolo 636

NITROGEN-FIXATION Inhibition of symbiotic nitrogen-fixation by gallic and g 57

NITROGEN-FIXING Inhibition of nitrogen-fixing and nitrifying bacteria by se 603
NITROGEN-FIXING Inhibition of nitrogen-fixing and nitrifying bacteria by se 604

NODULATION Inhibition of nodulation of inoculated legumes by pioneer plant 607

NOMENCLATURE Nomenclature of the genus Achillea. In: National list of scien 778

NON-ALLERGENIC Non-allergenic native and cultivated plants in Oklahoma: Lan 400

NOXIOUS Field surveys for noxious weeds in Oklahoma, Project 12-16-5-2301. 477

NUTRIENT Nutrient and biomass allocation strategies for five grass species 8
NUTRIENT Nutrient and biomass allocation in five grass species in an Oklaho 11
NUTRIENT Standing state and cycling of nitrogen in soil-vegetation componen 61
NUTRIENT Micronutrients in soils of the tall-grass prairie of Oklahoma. Pro 120
NUTRIENT Hydrologic and nutrient relationships in a pine-hardwood forest in 137
NUTRIENT Effects of chemical composition of soil on micronutrients and othe 298
NUTRIENT Nutrient analysis of selected forbs on clearcut areas in southeast 594

NUTRITION Seed dispersal and mineral nutrition in succession in abandoned f 611

NUTTALL On Nuttall's trail through Arkansas. J. Arnold Arb. 8: 24-52. {SE N 528

OAK The short grass plains and post oak-blackjack woodland of Oklahoma in h 56
OAK Comparative studies of the animal communities of oak-hickory forests in 76
OAK Toxic plants of Oklahoma: Oak, black locust, Kentucky coffee-tree, and 96
OAK The changing phytosociology of the oak-hickory forest of the Ozark Moun 117
OAK An evaluation of the quadrat method in the blackjack-post oak forest. P 152
OAK Effects of various plant and environmental factors on post oak response 159
OAK Vegetation response following winged elm (Ulmus alata Michx.) and oak (172
OAK Vegetational responses following winged elm and oak control in Oklahoma 173
OAK Description and dynamics of some western oak forests in Oklahoma. Ph. D 191
OAK Ordination and classification of western oak forests in Oklahoma. Amer. 192
OAK A comparison of post oak-blackjack oak communities in two major soil ty 199
OAK Geographic distribution of two species of oaks in Oklahoma in relation 284
OAK Drought years in central Oklahoma from 1710 to 1959 calculated from ann 295
OAK A study of interspecific hybridization in a natural oak population with 315
OAK Interspecific hybridization in a natural oak population with particular 316
OAK Biomass, annual net primary production and dynamics of six mineral elem 345
OAK Correlation analysis of rainfall and annual ring index of central Oklah 350
OAK Biomass, annual net primary production, and dynamics of six mineral ele 351
OAK A quantitative comparison between an oak forest and an oak savannah in 352
OAK Energy analyses of forestry practices in oak-pine forests and loblolly 383
OAK A study of the environment and floristic composition of an oak-hickory 426
OAK Shinnery oak can be controlled. Proc. Southern Weed Conf. 9: 95-98. {SW 466
OAK Progress in shinnery oak and sand sage control at Woodward. Prog. Rept. 467
OAK A summary of fire and forage research on shinnery oak rangelands. Proc. 468
OAK Competitive and allelopathic suppression of understory by Oklahoma oak 476
OAK Prairie dogs in shinnery (oak scrub) savannah. Ecology 23: 110-115. {NW 522
OAK The composition of a post oak forest in south-central Oklahoma. Southw. 549
OAK Oak (Quercus spp.) sprout growth rates on a central Oklahoma shallow sa 586
OAK An evaluation of the variable-radius and paired-tree methods in the bla 619
OAK A comparison of the predominant soil microflora in eight vegetative com 706
OAK Preliminary ecological study of the shinnery oak area of western Oklaho 842

OAK-HICKORY Comparative studies of the animal communities of oak-hickory fo 76
OAK-HICKORY The changing phytosociology of the oak-hickory forest of the Oz 117
OAK-HICKORY A study of the environment and floristic composition of an oak- 426
OAK-HICKORY A comparison of the predominant soil microflora in eight vegeta 706

OAK-PINE Energy analyses of forestry practices in oak-pine forests and lobl 383

OAK-PINE A comparison of the predominant soil microflora in eight vegetativ 706

OAT Effects of oat straw and native hay mulches on plant composition and ed 610

OLD-FIELD Inhibition of symbiotic nitrogen-fixation by gallic and gallotann 57
OLD-FIELD Acetylene reduction (nitrogen fixation) and old-field succession 359
OLD-FIELD Acetylene reduction (N-fixation) of glucose amended soils from ce 361
OLD-FIELD Acetylene reduction (N-fixation) in soil and old-field succession 362
OLD-FIELD Inhibitional effects of Digitaria sanguinalis and possible role i 533
OLD-FIELD Inhibitional effects of Digitaria sanguinalis and possible role i 534
OLD-FIELD Effects of certain plants of old-field succession on the growth o 538
OLD-FIELD Some aspects of structure and function in Oklahoma old-field succ 562
OLD-FIELD Some aspects of structure and function in Oklahoma old-field succ 563
OLD-FIELD Pollination of Cnidoscolus texanus (Euphorbiaceae) in south centr 564
OLD-FIELD Causes of decreases in residual carbohydrase activity in soil dur 612
OLD-FIELD Allelopathy as expressed by Helianthus annuus, and its role in ol 849
OLD-FIELD Allelopathy as expressed by Helianthus annuus and its role in old 850

ORCHIDACEAE Listera australis Lindley (Orchidaceae) new for Oklahoma. South 431

ORCHIDS The tallgrass prairie orchids Platanthera leucophaea (Nutt.) Lindl. 70
ORCHIDS The native orchids of the prairies and plains region of North Ameri 430
ORCHIDS New native orchids for Oklahoma and Kansas. Proc. Okla. Acad. Sci. 432
ORCHIDS Orchids and other new and interesting plants from Oklahoma. Pub. No 434
ORCHIDS The prairie fringed orchids: A pollinator-isolated species pair. Rh 681

ORDINATION Ordination and classification of mature bottomland forests in no 140
ORDINATION Description and dynamics of some western oak forests in Oklahoma 191
ORDINATION Ordination and classification of western oak forests in Oklahoma 192
ORDINATION Revegetation of unreclaimed coal strip mines in Oklahoma. II. Pl 256
ORDINATION Woody vegetation of southeastern LeFlore County, Oklahoma, in re 347
ORDINATION An analysis of the vegetation of buffalo wallows in relation to 579
ORDINATION The relationship of plant species heterogeneity to soil variatio 580
ORDINATION Relationships of vegetation and environment in buffalo wallows. 581

ORGANIC A comparison of organic carbon, hydrogen ion concentration, and vol 368

ORIGIN Silverleaf nightshade (Solanum eleagnifolium) origin, distribution, 71
ORIGIN Origin and distribution of polyploid Achillea (Compositae) in wester 774

ORNAMENTAL Toxic plants of Oklahoma: Ornamental houseplants. J. Okla. Vet. 99
ORNAMENTAL Toxic plants of Oklahoma: Ornamental shrubs. J. Okla. Vet. Med. 101
ORNAMENTAL Ornamental plants potentially hazardous to cattle. Bovine Pract. 105
ORNAMENTAL The ecological basis of ornamental gardening in Oklahoma. M. S. 501

ORPHAN A comparison of forest vegetation on orphan strip mines and adjacent 299

OUACHITA Geology of the southern Ouachita Mountains of Oklahoma. Okla. Geol 324
OUACHITA Deer browse in the Ouachita forest in Oklahoma. J. Wildl. Manage. 658
OUACHITA Hedyotis ouachitana (Rubiaceae): a new species from the Ouachita M 698

OVERGRAZED Insect distribution and seasonal succession in overgrazed and no 157
OVERGRAZED The distribution of rodents in overgrazed and normal grasslands 575
OVERGRAZED Rodent distribution in overgrazed and normal grasslands. M. S. T 576

OVERGRAZING Overgrazing from an ecological point of view. Proc. Okla. Acad. 78
OVERGRAZING The effect of overgrazing and erosion upon the biota of the mix 696
OVERGRAZING The effect of overgrazing on insect populations. Proc. Okla. Ac 836

OVERSTORY An analysis of selected Oklahoma upland forest stands including b 375

OZARK The changing phytosociology of the oak-hickory forest of the Ozark Mo 117
OZARK A short phytogeography of the prairies and Great Plains of central No 646
OZARK Additions to the Oklahoma flora from the Oklahoma Ozarks. Proc. Okla. 803
OZARK Vascular plants of the Oklahoma Ozarks. Ph. D. Diss., Okla. State Uni 804
OZARK Some geographic relationships of the vascular flora of the Oklahoma O 825
OZARK A list of the vascular flora of the Oklahoma Ozarks. Proc. Okla. Acad 826

PAIRED-TREE An evaluation of the variable-radius and paired-tree methods in 619

PANAX Panax quinquefolium (Araliaceae) in southeastern Oklahoma. Rhodora 87 403

PANICUM Cytological and morphological variation in Panicum virgatum L. M. S 80
PANICUM The physiological ecology of upland and lowland Panicum virgatum. P 505
PANICUM A comparative study of upland and bottomland types of Panicum virga 582
PANICUM An analysis of variation between upland and lowland switchgrass, Pa 583
PANICUM Carbon dioxide exchange characteristics of some prairie grass seedl 630

PARAQUAT Interseeding and paraquat effects on central and eastern rangeland 224

PASPALUM Pollination systems in Paspalum dilatatum Poir. (Poaceae): An exam 10

PASTURE Vegetation-soil relationships in Flint Hills bluestem pastures. J. 27
PASTURE Pasture and range plants. Phillips Petrol. Co., Bartlesville, OK. { 29
PASTURE A partial inventory of insect populations in tallgrass prairie past 81
PASTURE Brush control and pasture development in the Red Plains. Agron. J. 208
PASTURE The grasses and grasslands of Oklahoma. Status of current research 219
PASTURE Carrying capacity of pastures in Muskogee County, Oklahoma. M. S. T 231
PASTURE Effect of fertilization on native grass pastures in Oklahoma. J. Ra 331
PASTURE Vegetation and soil conditions of prairie and pasture plots in cent 367
PASTURE Energy analysis of Oklahoma rangelands and improved pastures. J. Ra 384
PASTURE Range and pasture studies and grazing recommendations for the south 474
PASTURE A comparison of the insects of a virgin prairie, a grazed pasture, 537
PASTURE Some effects of winter burning on a moderately grazed pasture. Ecol 560
PASTURE Establishment and management of tame and native grass pastures in O 587
PASTURE Establishment of warm and cool season grass pastures on wooded site 664

PATTERN Allelopathic effects of Platanus occidentalis L. in relation to the 17
PATTERN Flowering patterns and production in a central Oklahoma grassland. 28
PATTERN Weather patterns and ragweed production in Tulsa, Oklahoma. Ann. Al 87
PATTERN Ragweed pollen and weather patterns in Tulsa, Oklahoma. Bull. Ecol. 88
PATTERN An assessment of weather patterns and ragweed pollen in Tulsa over 89
PATTERN Evaluation of Payne County, Oklahoma farmlands and vegetation patte 177
PATTERN Wind patterns and the seasonal occurrence of Juniperus pollen in Tu 402
PATTERN Allelopathic effects of Celtis laevigata Willd. as related to patte 417
PATTERN Possible role of Ambrosia psilostachya on patterning and succession 495
PATTERN Possible role of Ambrosia psilostachya on pattern and succession in 496
PATTERN Allelopathic effects of Sporobolus pyramidatus on vegetational patt 589
PATTERN Allelopathic effects of Sporobolus pyramidatus on vegetational patt 590
PATTERN The ecology of Honey Creek: Temporal patterns of the travertine per 596
PATTERN Factor analysis of tree distribution patterns in Oklahoma. Ecology 656
PATTERN Shrub distribution patterns in Oklahoma with comparisons to those e 657
PATTERN The pattern of retrogression of native vegetation in north central 686
PATTERN Pattern of retrogression of native vegetation in north central Okla 688

PERIDINIOPSIS Studies in the dinoflagellate genera, Peridinium and Peridini 321

PERIDINIUM Studies in the dinoflagellate genera, Peridinium and Peridiniops 321

PERIPHYTON Spatial and temporal variation in productivity, species diversit 150
PERIPHYTON Species composition and diversity of periphyton in the Grand Riv 571
PERIPHYTON The ecology of Honey Creek: Temporal patterns of the travertine 596
PERIPHYTON Variation with stream order in species composition, diversity, b 665

PERMIAN Water fluctuation as a factor in the life of the higher plants of a 181

PERSIMMON The occurrence of persimmon and buckeye in the Wichita Mountains 285

PETROLEUM Petroleum and surface vegetation. Proc. Okla. Acad. Sci. 10: 110- 267

PETROLOGY Geology and petrology of the Wichita Mountains. Okla. Geol. Surv. 319

PHENOLOGICAL Phenological notes for 1928 on middle western spring flowers. 269
PHENOLOGICAL A phenological comparison of mowed and unmowed tallgrass 488

PHENOLOGY Phenology of a native tall-grass prairie in central Oklahoma. Eco 13
PHENOLOGY Phenology of a native tall-grass prairie in central Oklahoma. M. 14
PHENOLOGY Flowering patterns and production in a central Oklahoma grassland 28
PHENOLOGY Notes on the phenology of middle western spring flowers. Proc. Ok 268
PHENOLOGY Phenological notes for 1928 on middle western spring flowers. Pro 269
PHENOLOGY A preliminary report on the seasonal aspects of six habitats near 308
PHENOLOGY Oklahoma spring flowers in autumn. Proc. Okla. Acad. Sci. 17: 60- 325
PHENOLOGY The quantitative phenology of two plant communities in Osage Coun 355
PHENOLOGY A phenological comparison of mowed and unmowed tallgrass prairie 488
PHENOLOGY The phenology of woody plants in Norman, Oklahoma during the spri 544
PHENOLOGY Phenology of spring weeds in Norman, Oklahoma. Southw. Natur. 1: 546
PHENOLOGY Growth and development of five species of range grasses in centra 598

PHENOTYPIC Phenotypic variation and relationships of several variables in s 458

PHLOX Distribution, habitat, and reproductive biology of Phlox oklahomensis 703
PHLOX Chromosome number of Phlox oklahomensis Wherry; IOPB Chromoso 704
PHLOX Phlox oklahomensis. Status report. Endangered Species Office, U. S. F 747
PHLOX Phlox pilosa var. longipilosa. Status report. Endangered Species Offi 752
PHLOX Phlox pilosa L. var. longipilosa – A new combination. Sida 9: 183. { 754

PHOSPHORUS Calcium, magnesium, phosphorus, and potassium budgets of a tall 674
PHOSPHORUS A model for phosphorus and potassium flux in a tall-grass prairi 675
PHOSPHORUS Plant development and sequence in artificial ponds with and with 769

PHOSPHORUS-ENRICHED Plant development and sequence in artificial ponds with 769

PHOTOSENSITIZING Toxic plants of Oklahoma: Photosensitizing plants and vetc 103

PHREATOPHYTIC Water relations of some grasses with phreatophytic properties 205

PHYLLOSTACHYAE A new species of Carex (Section Phyllostachyae) from 821

PHYSALIS Systematic studies in Physalis (Solanaceae). Ph. D. Thesis, Univ. 716

PHYSICO-CHEMICAL The ecology of Honey Creek: Temporal patterns of the trave 596

PHYSIOGRAPHIC Biotic and physiographic succession on abandoned eroded farml 695
PHYSIOGRAPHIC Biotic and physiographic succession on abandoned eroded farml 697

PHYSIOLOGICAL Some morphological and physiological effects of density and d 197
PHYSIOLOGICAL The physiological ecology of upland and lowland Panicum virga 505

PHYTODERMATOSES Toxic plants of Oklahoma: Species causing phytodermatoses. 94

PHYTOGEOGRAPHY A short phytogeography of the prairies and Great Plains of c 646

PHYTOPLANKTON Water quality and phytoplankton communities in Lake Thunder 47
PHYTOPLANKTON Species composition and diversity of phytoplankton in the Gra 570

PHYTOSOCIOLOGICAL A phytosociological analysis of the tall-grass prairie in 592
PHYTOSOCIOLOGICAL A phytosociological analysis of the tall-grass prairie in 593
PHYTOSOCIOLOGICAL Phytosociological analysis of a tall-grass prairie in Mar 599
PHYTOSOCIOLOGICAL Phytosociological analysis of Oklahoma upland forest spec 633

PHYTOSOCIOLOGY The woody vegetation of Bear's Glen, a Washington Irving sto 40
PHYTOSOCIOLOGY The changing phytosociology of the oak-hickory forest of the 117
PHYTOSOCIOLOGY Ordination and classification of mature bottomland forests i 140
PHYTOSOCIOLOGY Succession in grasslands: thirty-two years of change in a ce 141
PHYTOSOCIOLOGY Description and dynamics of some western oak forests in Okla 191
PHYTOSOCIOLOGY Ordination and classification of western oak forests in Okla 192
PHYTOSOCIOLOGY Revegetation of unreclaimed coal strip mines in Oklahoma. II 256
PHYTOSOCIOLOGY Woody vegetation of southeastern LeFlore County, Oklahoma 347
PHYTOSOCIOLOGY The vegetation of Muskogee County, Oklahoma. Amer. Midl. 409
PHYTOSOCIOLOGY The vegetation of the Caddo County Canyons, Oklahoma. Ec 410
PHYTOSOCIOLOGY Ecological factors affecting the distribution of woody veget 420
PHYTOSOCIOLOGY A study of the environment and floristic composition of an o 426
PHYTOSOCIOLOGY A vegetation analysis of a pimpled prairie in northeastern O 491
PHYTOSOCIOLOGY An analysis of an elm-ash floodplain community near Norman, 541
PHYTOSOCIOLOGY Some aspects of structure and function in Oklahoma old-field 562
PHYTOSOCIOLOGY A study of the role of Rhus copallina in the community dynam 565
PHYTOSOCIOLOGY The role of Rhus copallina in the dynamics of the forest-pra 566
PHYTOSOCIOLOGY A phytosociological analysis of the tall-grass prairie in no 592
PHYTOSOCIOLOGY Correlation between vegetation and geologic formations in Ok 597
PHYTOSOCIOLOGY A statistical method for determining quadrat size and adequa 606
PHYTOSOCIOLOGY The species-area curve. Ecology 36: 7-11. {methods phytosoci 614
PHYTOSOCIOLOGY Diversity in tree species in Oklahoma upland forests. Ecolog 632
PHYTOSOCIOLOGY Factor analysis of tree distribution patterns in Oklahoma. E 656
PHYTOSOCIOLOGY Shrub distribution patterns in Oklahoma with comparisons to 657
PHYTOSOCIOLOGY Dispersion and co-dispersion in Oklahoma upland forests. Bul 678

PIGMENT Spatial and temporal variation in productivity, species diversity, 150

PIMPLE Ecological interpretation for small mounds in landscapes of eastern 24
PIMPLE A vegetation analysis of a pimpled prairie in northeastern Oklahoma. 491

PINE Introgressive hybridization between shortleaf and loblolly pine in sou 1
PINE Hydrologic and nutrient relationships in a pine-hardwood forest in sou 137
PINE Plant associations as site predictors in the pine-hardwood tension zon 187
PINE Herbicides for release of shortleaf pine and native grass. Weeds 15: 1 207
PINE Bio-economic analysis of even aged shortleaf pine stands in southeaste 211
PINE Natural variation in rooting ability of western provenances of shortle 223
PINE Studies in the cembroid pines. M. S. Thesis, Univ. Okla., Norman. {PH 237
PINE Energy analyses of forestry practices in oak-pine forests and loblolly 383
PINE Phenotypic variation and relationships of several variables in shortle 458
PINE Eleven year biologic-economic evaluation of undesirable hardwood sites 489
PINE Relationship of soil properties to site index of shortleaf pine and di 526
PINE Natural regeneration of loblolly pine 230 miles northwest of its norma 584
PINE A comparison of the predominant soil microflora in eight vegetative co 706
PINE Shortleaf pine (Pinus echinata) in Bryan County, Oklahoma. Southw. Nat 728

PINE Stem frequency distribution and sampling of the hardwood component of 760
PINE Shortleaf pine site index – Soils and plant associations on the coast 851
PINE Environmental parameters and site identification for potential expansi 853

PINE-HARDWOOD Hydrologic and nutrient relationships in a pine-hardwood fore 137
PINE-HARDWOOD Plant associations as site predictors in the pine-hardwood te 187
PINE-HARDWOOD Stem frequency distribution and sampling of the hardwood comp 760

PINUS Herbicides for release of shortleaf pine and native grass. Weeds 15: 207
PINUS Natural regeneration of loblolly pine 230 miles northwest of its norm 584
PINUS Shortleaf pine (Pinus echinata) in Bryan County, Oklahoma. Southw. Na 728
PINUS Shortleaf pine site index – Soils and plant associations on the coas 851

PINYON The vegetation of the Mesa de Maya region of Colorado, New Mexico, a 639
PINYON Some botanical studies in the Black Mesa region of Oklahoma. Rhodora 640

PIONEER Algae as pioneers in plant succession and their importance in erosi 66
PIONEER Inhibition of nodulation of inoculated legumes by pioneer plant spe 607

PLAIN Vegetation of the Salt Plains Wildlife Refuge, Jet, Oklahoma. Ph. D. 33
PLAIN A floristic study of the vascular plants of the Gypsum Hills and Redb 35
PLAIN Floristic components of the Gypsum Hills and Redbed Plains of southwe 36
PLAIN The short grass plains and post oak-blackjack woodland of Oklahoma in 56
PLAIN The eastern boundary of the Great Plains in north central Oklahoma. P 63
PLAIN The effect of salinity on the distribution of algae at the Salt Plain 113
PLAIN Reproduction of woody plants in the absence of grazing. Proc. Okla. A 119
PLAIN Investigations in erosion control and reclamation of eroded land at t 174
PLAIN Brush control and pasture development in the Red Plains. Agron. J. 42 208
PLAIN Revegetation and gully-control experiments in the Red Plains Region. 209
PLAIN Relation of climatic conditions, soil characteristics, and tree devel 294
PLAIN Retrogression of native range vegetation in the sandy high plains. M. 317
PLAIN The role of the bison in maintaining the short grass plains. Ecology 393
PLAIN The native orchids of the prairies and plains region of North America 430
PLAIN New and interesting plants from the Great Plains. Rhodora 74: 378-388 463
PLAIN Atlas of the flora of the Great Plains. Iowa State Univ. Press, Ames. 464
PLAIN Flora of the Great Plains. Univ. Kansas Press, Lawrence. {flora tallg 465
PLAIN Spraying 2,4-D by airplane on sand sagebrush and other plants of the 469
PLAIN Fourteen year summary of range improvement studies. Bull. U. S. South 470
PLAIN Range and pasture studies and grazing recommendations for the souther 474
PLAIN The Gulf Coastal Plain Ultisol and Alfisol ecosystem: surface geology 502
PLAIN Revegetation of eroded lands in the High Plains of Texas and Oklahoma 508
PLAIN The ecology of the western Oklahoma salt plains. Pub. Univ. Okla. Bio 519
PLAIN A short phytogeography of the prairies and Great Plains of central No 646
PLAIN Methods of reestablishing buffalo grass on cultivated land in the Gre 647
PLAIN Drought survival of native grass species in the central and southern 648
PLAIN The Southern Great Plains. Grass: USDA Yearbook of Agric., p. 503-540 649
PLAIN Natural revegetation of abandoned farmland in the central and souther 650
PLAIN Natural vegetation as an indicator of the capabilities of land for cr 667
PLAIN The natural vegetation of the Great Plains region. Ann. Assoc. Amer. 668
PLAIN The relation of forest vegetation to soils and geology in the gulf co 729
PLAIN Species-soil relationships in the Great Salt Plains of northern Oklah 792
PLAIN Progress of regrassing in the Southern Great Plains. Soil Cons. 5: 57 829
PLAIN Grasslands of the Great Plains. Johnsen Pub. Co., Lincoln, Neb. {prai 831
PLAIN New and interesting plants from the Great Plains. Southw. Natur. 18: 834
PLAIN Shortleaf pine site index – Soils and plant associations on the coas 851

PLANKTON Water quality and phytoplankton communities in Lake Thunderbird, C 47
PLANKTON Plankton production and species distribution in the limnological p 379
PLANKTON Species composition and diversity of phytoplankton in the Grand Ri 570

PLANTATIONS Energy analyses of forestry practices in oak-pine forests and l 383

PLATANUS Allelopathic effects of Platanus occidentalis L. M. S. Thesis, Uni 16
PLATANUS Allelopathic effects of Platanus occidentalis L. in relation to th 17
PLATANUS Allelopathic effects of Platanus occidentalis. Bull. Torrey Bot. C 18
PLATANUS Growth and distribution of Platanus occidentalis in Oklahoma as in 30

PLATT Platt National Park: Environment and ecology. Univ. Okla. Press, Norm 42
PLATT The grasslands of Platt National Park, Oklahoma. Southw. Natur. 4: 45 170
PLATT Final report on vegetation and microenvironments of Platt National Pa 171

PLOWED A comparison of the insects of a virgin prairie, a grazed pasture, a 537
PLOWED Plant population changes in a native prairie plot plowed annually ov 558
PLOWED Plant succession and yield of living plant material in a plowed prai 618

PLOWING Effects of fencing and plowing on plant succession in a revegetatin 557

POACEAE Pollination systems in Paspalum dilatatum Poir. (Poaceae): An examp 10
POACEAE Systematic studies in the Poaceae and Cyperaceae. Ph. D. Thesis, Un 73
POACEAE Hybridization and introgression between Elymus canadensis and Elymu 498
POACEAE Systematics of Aristida dichotoma, basiramea, and curtisii (Poaceae 800
POACEAE Studies in the Danthonieae (Poaceae). Ph. D. Thesis, Univ. Okla., N 854

PODOPHYLLUM Podophyllum peltatum f. deamii from Bryan County, Oklahoma. 724

POINT-CENTERED A modification of the point-centered quarter method for gras 550

POISON Toxic plants of Oklahoma: Species causing phytodermatoses. J. Okla. 94
POISON Toxic plants of Oklahoma: Locoweeds. J. Okla. Vet. Med. Assoc. 33: 1 95
POISON Toxic plants of Oklahoma: Oak, black locust, Kentucky coffee-tree, a 96
POISON Toxic plants of Oklahoma: Thornapples and nightshades. J. Okla. Vet. 97
POISON Toxic plants of Oklahoma: Bitterweeds and sneezeweeds. J. Okla. Vet. 98
POISON Toxic plants of Oklahoma: Ornamental houseplants. J. Okla. Vet. Med. 99
POISON Toxic plants of Oklahoma: Hemlocks. J. Okla. Vet. Med. Assoc. 34: 30 100
POISON Toxic plants of Oklahoma: Ornamental shrubs. J. Okla. Vet. Med. Asso 101
POISON Toxic plants of Oklahoma: Coffeeweeds and sennas. J. Okla. Vet. 102
POISON Toxic plants of Oklahoma: Photosensitizing plants and vetch. J. Okla 103
POISON Toxic plants of Oklahoma: Groundsels and rattlebox. J. Okla. Vet. Me 104
POISON Ornamental plants potentially hazardous to cattle. Bovine Pract. 18: 105
POISON Toxic plants of Oklahoma: Ferns and horsetails. J. Okla. Vet. Med. A 106
POISON Toxic plants of Oklahoma: Cyanogenic plants. J. Okla. Vet. Med. Asso 107
POISON Toxic plants of Oklahoma: Milkweeds. J. Okla. Vet. Med. Assoc. 36: 7 108
POISON Toxic plants of Oklahoma: Species causing reproductive disorders. J. 109
POISON Toxic plants of Oklahoma: Species causing acute respiratory distress 110
POISON Toxic plants of Oklahoma: Cocklebur. J. Okla. Vet. Med. Assoc. 37: 6 111
POISON Toxic plants of Oklahoma: Cotton and cottonseed. J. Okla. Vet. Med. 202
POISON Toxic plants of Oklahoma: Clovers. J. Okla. Vet. Med. Assoc. 36: 30- 203
POISON Potentially poisonous or otherwise harmful higher plants of Oklahoma 287
POISON Plants of Oklahoma and Texas capable of producing cyanogenic poisons 685

POLEMONIACEAE Distribution, habitat, and reproductive biology of Phlox okla 703
POLEMONIACEAE Gathering of Ipomopsis rubra (Polemoniaceae) nectar by Bom 775

POLLEN Ragweed pollen and weather patterns in Tulsa, Oklahoma. Bull. Ecol. 88

POLLEN An assessment of weather patterns and ragweed pollen in Tulsa over a 89
POLLEN A limited-range provenance test of open-pollenated progeny of Carya 300
POLLEN Wind patterns and the seasonal occurrence of Juniperus pollen in Tul 402

POLLINATION Pollination systems in Paspalum dilatatum Poir. (Poaceae): An e 10
POLLINATION Comparative pollination ecology of Pyrrhopappus geiseri and Pyr 38
POLLINATION The pollination ecology of Solanum rostratum (Solanaceae). Amer 69
POLLINATION Pollination biology of Polygala alba Nutt. (Polygalaceae). Ph. 445
POLLINATION Pollination of Cnidoscolus texanus (Euphorbiaceae) in south cen 564

POLLUTION The effect of cement-kiln dust on vegetation. M. S. Thesis, Univ. 21
POLLUTION Chemical and biological analysis of a hazardous waste disposal si 360

POLLINATION Pollination biology of Polygala alba Nutt. (Polygalaceae). Ph. D. 445

POLYGALACEAE Pollination biology of Polygala alba Nutt. (Polygalaceae). Ph. 445

POLYGONACEAE A taxonomic investigation of the Polygonum pensylvanicum L. (P 201
POLYGONACEAE A taxonomic study of the subgenus Pterogonum of the genus Erio 311

POLYGONUM Allelopathic effects of Polygonum aviculare L. Ph. D. Thesis, Uni 19
POLYGONUM A taxonomic investigation of the Polygonum pensylvanicum L. (Poly 201

POLYPLOID Origin and distribution of polyploid Achillea (Compositae) in wes 774

POND The relation of stock pond vegetation to geological formations in Mars 145
POND Changes in vegetation on stock pond dams near Norman, Oklahoma. M. S. 366
POND The vegetation of stock pond dams in central Oklahoma. Amer. Midl. Nat 371
POND Epiphytic diatoms as indicators of water quality in farm and experimen 765
POND Survey of algae in farm ponds in Bryan County, Oklahoma. Southw. Natur 766
POND Epiphytic diatoms in farm ponds and experimental ponds in Bryan County 767
POND Epiphytic diatoms as indicators of water quality in farm and experimen 768
POND Plant development and sequence in artificial ponds with and without ph 769
POND A comparison of algal succession between farm ponds with sandstone and 856
POND A comparison of algal composition and seasonal succession between pond 857

PONDEROSA The vegetation of the Mesa de Maya region of Colorado, New Mexico 639
PONDEROSA Some botanical studies in the Black Mesa region of Oklahoma. Rhod 640

POPULATION A partial inventory of insect populations in tallgrass prairie p 81
POPULATION Low intensity forest sampling through use of stem frequency dist 167
POPULATION Rodent populations and their reactions to grazing intensities on 242
POPULATION Relationships of sugar maples (Acer saccharum and A. grandidenta 252
POPULATION The influence of habitat quality on density of cotton rat popula 258
POPULATION A population study of Chara zeylanica in Texas, Oklahoma, and Ne 275
POPULATION Differential selection in juniper populations from the Baum Lime 282
POPULATION A study of interspecific hybridization in a natural oak populati 315
POPULATION Interspecific hybridization in a natural oak population with par 316
POPULATION Populations and range effects of rodents on the sand sagebrush g 457
POPULATION Condensed tannins in soil: inputs and effects on microbial popul 486
POPULATION Replacement of a population of Johnson grass by a vine-forb comm 556
POPULATION Plant population changes in a native prairie plot plowed annuall 558
POPULATION The effect of overgrazing on insect populations. Proc. Okla. Aca 836

POPULUS Genetic variation in juvenile characteristics of Populus deltoides 497

POST OAK-BLACKJACK The short grass plains and post oak-blackjack woodland o 56
POST OAK-BLACKJACK A comparison of post oak-blackjack oak communities in tw 199
POST OAK-BLACKJACK Biomass, annual net primary production and dynamics of s 345

POST OAK-BLACKJACK Biomass, annual net primary production, and dynamics of 351

POTASSIUM Calcium, magnesium, phosphorus, and potassium budgets of a tall g 674
POTASSIUM A model for phosphorus and potassium flux in a tall-grass prairie 675

PRAIRIE Nutrient and biomass allocation strategies for five grass species i 8
PRAIRIE Differential response of woody and herbaceous species to summer and 9
PRAIRIE Nutrient and biomass allocation in five grass species in an Oklahom 11
PRAIRIE Phenology of a native tall-grass prairie in central Oklahoma. Ecolo 13
PRAIRIE Phenology of a native tall-grass prairie in central Oklahoma. M. S. 14
PRAIRIE Injury and death or recovery of trees in a prairie climate. Ecol. M 20
PRAIRIE Tall grass defeats prairie dogs. Soil Cons. 20:103-105. {SW grassla 22
PRAIRIE Distribution and control of several woody plants in Texas and Oklah 25
PRAIRIE Vegetation-soil relationships in Flint Hills bluestem pastures. J. 27
PRAIRIE Flowering patterns and production in a central Oklahoma grassland. 28
PRAIRIE Pasture and range plants. Phillips Petrol. Co., Bartlesville, OK. { 29
PRAIRIE A preliminary study of buffalo wallows in the vicinity of Norman, O 45
PRAIRIE Evaluation of sampling techniques on tall-grass prairie. J. Range M 48
PRAIRIE The short grass plains and post oak-blackjack woodland of Oklahoma 56
PRAIRIE Standing state and cycling of nitrogen in soil-vegetation component 61
PRAIRIE The tallgrass prairie orchids Platanthera leucophaea (Nutt.) Lindl. 70
PRAIRIE Analysis of morphological variation in Schizachyrium scoparium (lit 77
PRAIRIE Overgrazing from an ecological point of view. Proc. Okla. Acad. Sci 78
PRAIRIE The vegetation of Oklahoma. Ecol. Monogr. 1: 99-188. {forest tree g 79
PRAIRIE A partial inventory of insect populations in tallgrass prairie past 81
PRAIRIE Microclimates of three grassland plots in central Oklahoma. M. S. T 82
PRAIRIE A survey of the tall grass prairies in northeastern Oklahoma. M. S. 83
PRAIRIE A survey of the tall-grass prairie in northeastern Oklahoma. Southw 91
PRAIRIE Micronutrients in soils of the tall-grass prairie of Oklahoma. Proc 120
PRAIRIE Seasonal aspects in certain prairie and savannah communities. M. S. 121
PRAIRIE Some observations on the number and size of samples necessary to ma 122
PRAIRIE Plant distribution in a central Oklahoma prairie. Proc. Okla. Acad. 123
PRAIRIE Fluctuations in biotic communities. III. Aspection in a ravine sere 124
PRAIRIE Fluctuations in biotic communities. V. Aspection in mixed-grass pra 125
PRAIRIE Ecology of the grassland biome. M. S. Thesis, Univ. Okla., Norman. 126
PRAIRIE The grassland biome. Ecol. Monogr. 10: 617-684. {prairie shortgrass 127
PRAIRIE A study of the native grasses of Comanche County, Oklahoma. M. S. T 130
PRAIRIE Interaction of disturbances in tallgrass prairie: a field experimen 139
PRAIRIE Succession in grasslands: thirty-two years of change in a central O 141
PRAIRIE The effect of early spring burning on vegetation in buffalo wallows 142
PRAIRIE Effects of disturbance on diversity in mixed-grass prairie. Vegetat 144
PRAIRIE Vegetation structure in a tallgrass prairie. M. S. Thesis, Univ. Ok 146
PRAIRIE Canopy structure of a tall-grass prairie. J. Range Manage. 27: 313- 147
PRAIRIE Insect distribution and seasonal succession in overgrazed and norma 157
PRAIRIE Preliminary studies in the evaluation of grassland sampling techniq 161
PRAIRIE Grassland communities of the Wichita Mountains Wildlife Refuge. Ph. 162
PRAIRIE Influence of soils and parent materials on grasslands of the Wichit 163
PRAIRIE Effect of mowing on a relict tall grass prairie. Proc. Okla. Acad. 164
PRAIRIE The grasslands of Platt National Park, Oklahoma. Southw. Natur. 4: 170
PRAIRIE The black-tailed prairie dog (Cynomys ludovicianus) as a member of 176
PRAIRIE The vegetation of the Wichita Mountains Wildlife Refuge. M. S. Thes 186
PRAIRIE Brush control and the welfare of lesser prairie chickens in western 188
PRAIRIE Effects on lesser prairie chickens of brush control in western Okla 189
PRAIRIE Identifying the prairie-forest transition using remote sensing. M. 196

PRAIRIE An annotated plant list for Adams Ranch. M. S. Thesis, Ft. Hays Kan 198
PRAIRIE The savannah concept and its use. Ecology 38: 435-442. {classificat 200
PRAIRIE Influence of grazing on plant succession of rangelands. Bot. Rev. 2 204
PRAIRIE A comparison of the soil fungi of a tall-grass prairie and of an ab 212
PRAIRIE A quantitative comparison of the soil algae of central Oklahoma pra 213
PRAIRIE Grasses and grasslands: systematics and ecology. Univ. Okla. Press, 218
PRAIRIE The grasses and grasslands of Oklahoma. Status of current research 219
PRAIRIE Classification of Oklahoma grasses. Ann. Okla. Acad. Sci. 6: 141-16 220
PRAIRIE Effects of mowing on a native tall grass prairie in central Oklahom 247
PRAIRIE A vegetational analysis of the Phillips Agricultural Demonstration 248
PRAIRIE Why prairies are treeless. Science N. S. 22: 55-56. {fire burning g 249
PRAIRIE The effects of nitrogen fertilization on native vegetation under cl 250
PRAIRIE The effects of one year's nitrogen fertilization on native vegetati 251
PRAIRIE Commerce of the prairies: or, The journal of a Santa Fe trader, dur 274
PRAIRIE Water stress of tallgrass prairie plants in central Oklahoma. J. Ra 277
PRAIRIE A comparison of soil microfungi of two grassland areas in central O 288
PRAIRIE Production characteristics of Oklahoma forages: Native range. Okla. 291
PRAIRIE Effects of chemical composition of soil on micronutrients and other 298
PRAIRIE The identification of some of the more common native Oklahoma grass 301
PRAIRIE A floristic study of the woody vegetation of the North American Cro 302
PRAIRIE Forage production, vegetative composition, and plant vigor in relat 304
PRAIRIE The claypan range site in northern Osage County, Oklahoma. J. Range 305
PRAIRIE Effect of grazing intensity on plant composition, vigor, and produc 306
PRAIRIE Aspection in a central Oklahoma tension zone area. M. S. Thesis, Un 320
PRAIRIE Developmental growth of four species of range grasses in north-cent 334
PRAIRIE Vegetation in relation to slope exposure and geology in the Arbuckl 335
PRAIRIE Mechanisms of secondary succession in a tall grass prairie. Ph. D. 336
PRAIRIE Changes in species composition of grassland communities in response 337
PRAIRIE A tour of the prairies. In crayon miscellany. Carey, Lea, and Blanc 338
PRAIRIE A tour on the prairies. Univ. Okla. Press, Norman. {presettlement} 339
PRAIRIE A quantitative comparison between an oak forest and an oak savannah 352
PRAIRIE Habitat preferences among representative wintering and breeding bir 353
PRAIRIE The life-form concept applied to prairie chicken habitat in Oklahom 354
PRAIRIE Identification and analysis of lesser and greater prairie chicken h 356
PRAIRIE Habitat used by lesser prairie chickens for feeding related to seas 357
PRAIRIE Acetylene reduction (N-fixation) in soil and old-field succession i 362
PRAIRIE Symbiotic and asymbiotic N2-fixation in a tall grass prairie. Soil 363
PRAIRIE Vegetation and soil conditions of prairie and pasture plots in cent 367
PRAIRIE Effects of moderate grazing on the composition and plant production 369
PRAIRIE Winter burning in central Oklahoma grassland. Ecology 38: 520-522. 370
PRAIRIE Restoration of eroded prairie with digested sludge. Proc. Okla. Aca 377
PRAIRIE Energy analysis of Oklahoma rangelands and improved pastures. J. Ra 384
PRAIRIE Prairie dogs, whitefaces, and blue grama. Wildl. Monogr. 3: 1-78. { 386
PRAIRIE Nitrogen fixation in selected prairie legumes as related to success 419
PRAIRIE The native orchids of the prairies and plains region of North Ameri 430
PRAIRIE Microbial decomposition of cellulose and native plant litter in a t 447
PRAIRIE A comparison of the soil fungi of a virgin prairie and an abandoned 461
PRAIRIE An investigation of the soil microflora of two grassland plots. Pro 462
PRAIRIE Fourteen year summary of range improvement studies. Bull. U. S. Sou 470
PRAIRIE The role of ecotypic variation in the distribution of the central g 473
PRAIRIE Range and pasture studies and grazing recommendations for the south 474
PRAIRIE A phenological comparison of mowed and unmowed tallgrass prairie si 488
PRAIRIE A vegetation analysis of a pimpled prairie in northeastern Oklahoma 491

PRAIRIE Range deterioration in south central Oklahoma. M. S. Thesis, Univ. 494
PRAIRIE The effect of disturbances in tall grass prairie sites on an index 503
PRAIRIE The physiological ecology of upland and lowland Panicum virgatum. P 505
PRAIRIE A study of soil compaction in a tall grass prairie. M. S. Thesis, U 506
PRAIRIE Prairie dogs in shinnery (oak scrub) savannah. Ecology 23: 110-115. 522
PRAIRIE Vegetation of an abandoned prairie dog town in tall-grass prairie. 523
PRAIRIE Ecological study of western Payne County, Oklahoma land utilization 535
PRAIRIE A comparison of the insects of a virgin prairie, a grazed pasture, 537
PRAIRIE Simulated impact of management practices upon the tallgrass prairie 540
PRAIRIE A modification of the point-centered quarter method for grassland a 550
PRAIRIE The relation of grazing to plant succession in the tall grass prair 551
PRAIRIE Effects of denudation on the productivity of grassland. Ecology 45: 552
PRAIRIE Influence of a wildfire in the Wichita Mountains Wildlife Refuge. E 554
PRAIRIE Plant population changes in a native prairie plot plowed annually o 558
PRAIRIE Some effects of winter burning on a moderately grazed pasture. Ecol 560
PRAIRIE A study of the role of Rhus copallina in the community dynamics of 565
PRAIRIE The role of Rhus copallina in the dynamics of the forest-prairie ec 566
PRAIRIE The distribution of rodents in overgrazed and normal grasslands in 575
PRAIRIE Rodent distribution in overgrazed and normal grasslands. M. S. Thes 576
PRAIRIE An analysis of the vegetation of buffalo wallows in relation to env 579
PRAIRIE The relationship of plant species heterogeneity to soil variation i 580
PRAIRIE Relationships of vegetation and environment in buffalo wallows. Ame 581
PRAIRIE An analysis of variation between upland and lowland switchgrass, Pa 583
PRAIRIE The vegetation regions of the prairie province. Bot. Gaz. 25: 381-3 585
PRAIRIE Oak (Quercus spp.) sprout growth rates on a central Oklahoma shallo 586
PRAIRIE Establishment and management of tame and native grass pastures in O 587
PRAIRIE A phytosociological analysis of the tall-grass prairie in northeast 592
PRAIRIE A phytosociological analysis of the tall-grass prairie in northeast 593
PRAIRIE Growth and development of five species of range grasses in central 598
PRAIRIE Phytosociological analysis of a tall-grass prairie in Marshall Coun 599
PRAIRIE Allelopathy and grassland improvement. In: Estes, J. R., and R. J. 608
PRAIRIE Causes of decreases in productivity in undisturbed tall grass prair 613
PRAIRIE Inhibition of nitrification by climax ecosystems. Amer. J. Bot. 59: 615
PRAIRIE Inhibition of nitrification by climax ecosystems. II. Additional ev 616
PRAIRIE Inhibition of nitrification by climax ecosystems. III. Inhibitors o 617
PRAIRIE Plant succession and yield of living plant material in a plowed pra 618
PRAIRIE Competitive responses of Bouteloua curtipendula (Michx.) Torr. Amer 624
PRAIRIE Preliminary producer data synthesis: 1970 comprehensive network sit 625
PRAIRIE Osage site, 1970 report: Primary production. USIBP Grassland Biome 626
PRAIRIE Plant community structure. In: N. R. French (ed) Preliminary analys 627
PRAIRIE A preliminary compartment model of a tall grass prairie: Osage 1970 628
PRAIRIE Systems analysis of a tall grass prairie. Proc. Third Midw. Prairie 629
PRAIRIE Carbon dioxide exchange characteristics of some prairie grass seedl 630
PRAIRIE Herbage dynamics of a tall grass prairie: Osage, 1971. USIBP Grassl 634
PRAIRIE Herbage dynamics of an Oklahoma tallgrass prairie: Osage 1972. USIB 635
PRAIRIE Ecosystem analysis of the tallgrass prairie: nitrogen cycle. Ecolog 636
PRAIRIE The true prairie ecosystem. Hutchinson Ross Publ. Co., Stroudsburg, 637
PRAIRIE A short phytogeography of the prairies and Great Plains of central 646
PRAIRIE Drought survival of native grass species in the central and souther 648
PRAIRIE The Southern Great Plains. Grass: USDA Yearbook of Agric., p. 503-5 649
PRAIRIE Natural revegetation of abandoned farmland in the central and south 650
PRAIRIE Observations on the grasslands of the central United States. Ohio S 651
PRAIRIE Possible mechanisms of the exclusion of Johnson grass by tall grass 660

PRAIRIE The natural vegetation of the Great Plains region. Ann. Assoc. Amer 668
PRAIRIE Calcium, magnesium, phosphorus, and potassium budgets of a tall gra 674
PRAIRIE A model for phosphorus and potassium flux in a tall-grass prairie. 675
PRAIRIE The prairie fringed orchids: A pollinator-isolated species pair. Rh 681
PRAIRIE The pattern of retrogression of native vegetation in north central 686
PRAIRIE The structure and function of ten western North American grasslands 687
PRAIRIE The structure and function of ten western North American grasslands 689
PRAIRIE The structure and function of ten North American grasslands. III. N 690
PRAIRIE The structure and function of ten western North American grasslands 691
PRAIRIE Some effects of differential clipping on six native grasses and on 692
PRAIRIE Effect of disturbance on microclimate and vegetation in tall grass 693
PRAIRIE The effect of overgrazing and erosion upon the biota of the mixed g 696
PRAIRIE A comparison of the predominant soil microflora in eight vegetative 706
PRAIRIE Allelopathic inhibition of weed invasion by Andropogon scoparius Mi 713
PRAIRIE The influence of burning on tillering and aboveground biomass parti 717
PRAIRIE A continuation of the investigation of the soil microflora of two g 718
PRAIRIE A broad analysis of fifteen sites in the tall-grass prairie of Okla 719
PRAIRIE The grassland communities of the Baum Limestone in Johnston County, 755
PRAIRIE The Chisholm Trail in geographic perspective. Proc. Okla. Acad. Sci 761
PRAIRIE Vertebrates in a prairie dog town. Proc. Okla. Acad. Sci. 50: 110-1 773
PRAIRIE Primary succession on granite outcrops in southwestern Oklahoma. Bu 797
PRAIRIE Notes on the growth and fruiting of Andropogon scoparius, with some 806
PRAIRIE North American prairie. Johnsen Pub. Co., Lincoln, Neb. {grassland} 830
PRAIRIE Grasslands of the Great Plains. Johnsen Pub. Co., Lincoln, Neb. {pr 831
PRAIRIE Effects of excessive mulch on development, yield, and structure of 832
PRAIRIE The effect of overgrazing on insect populations. Proc. Okla. Acad. 836
PRAIRIE A reconnaisance of a typical central Oklahoma prairie. M. S. Thesis 838
PRAIRIE Washington Irving on the prairie, or, a narrative of a tour in the 846
PRAIRIE A survey of the soil algae of two vegetation types in central Oklah 847
PRAIRIE An exploratory study on soil algae. Ecology 38: 309-313. {SC grassl 848

PRAIRIE-FOREST Identifying the prairie-forest transition using remote sensi 196

PRE-IMPOUNDMENT A pre-impoundment biological survey of Wister Reservoir, 279

PRIMARY Biomass, annual net primary production and dynamics of six mineral 345
PRIMARY Biomass, annual net primary production, and dynamics of six mineral 351
PRIMARY The primary productivity profile of Oklahoma. Okla. Biol. Survey Te 382
PRIMARY Primary production of vascular aquatic plants. Limnol. and Oceanog. 545
PRIMARY Osage site, 1970 report: Primary production. USIBP Grassland Biome 626
PRIMARY The structure and function of ten North American grasslands. III. N 690
PRIMARY The structure and function of ten western North American grasslands 691
PRIMARY Primary succession on granite outcrops in southwestern Oklahoma. Bu 797

PROBOSCIDEA The reproductive biology of Proboscidea louisianica (Martyniace 573
PROBOSCIDEA The reproductive biology of Proboscidea louisianica (Martyniace 574
PROBOSCIDEA Floral biology of Proboscidea louisianica (Martyniaceae). Rhodo 758
PROBOSCIDEA Chromosome numbers in Nama and Proboscidea; IOPB Chromos 780

PRODUCER Preliminary producer data synthesis: 1970 comprehensive network si 625
PRODUCER The structure and function of ten western North American grassland 691

PRODUCTION Flowering patterns and production in a central Oklahoma grasslan 28
PRODUCTION Weather patterns and ragweed production in Tulsa, Oklahoma. Ann. 87
PRODUCTION Reproduction of woody plants in the absence of grazing. Proc. Ok 119
PRODUCTION Vegetative habit and reproduction of Desmidium grevillii (Kütz) 154

PRODUCTION Production characteristics of Oklahoma forages: Native range. Ok 291
PRODUCTION Forage production, vegetative composition, and plant vigor in re 304
PRODUCTION Effect of grazing intensity on plant composition, vigor, and pro 306
PRODUCTION Biomass, annual net primary production and dynamics of six miner 345
PRODUCTION Biomass, annual net primary production, and dynamics of six mine 351
PRODUCTION Effects of moderate grazing on the composition and plant product 369
PRODUCTION Plankton production and species distribution in the limnological 379
PRODUCTION The primary productivity profile of Oklahoma. Okla. Biol. Survey 382
PRODUCTION Energy analyses of forestry practices in oak-pine forests and lo 383
PRODUCTION Energy analysis of Oklahoma rangelands and improved pastures. J. 384
PRODUCTION Net production relations of two tree species in southern Oklahom 424
PRODUCTION The significance of vegetative reproduction in Quercus. Madroño 490
PRODUCTION Primary production of vascular aquatic plants. Limnol. and Ocean 545
PRODUCTION Effects of denudation on the productivity of grassland. Ecology 552
PRODUCTION Plant succession and yield of living plant material in a plowed 618
PRODUCTION Osage site, 1970 report: Primary production. USIBP Grassland Bio 626
PRODUCTION Natural vegetation as an indicator of the capabilities of land f 667
PRODUCTION The structure and function of ten western North American grassla 687
PRODUCTION The structure and function of ten western North American grassla 689
PRODUCTION The structure and function of ten North American grasslands. III 690
PRODUCTION The structure and function of ten western North American grassla 691

PRODUCTIVITY Spatial and temporal variation in productivity, species divers 150
PRODUCTIVITY The primary productivity profile of Oklahoma. Okla. Biol. Surv 382
PRODUCTIVITY Effects of denudation on the productivity of grassland. Ecolog 552
PRODUCTIVITY Causes of decreases in productivity in undisturbed tall grass 613

PROGENY A limited-range provenance test of open-pollinated progeny of Carya 300

PROGRAM A soil improvement program for Oklahoma. Okla. A & M Coll. Circ. No 131
PROGRAM The Oklahoma Natural Heritage Program 1977-1984: a critical evaluat 480

PROVENANCE Natural variation in rooting ability of western provenances of s 223
PROVENANCE A limited-range provenance test of open-pollenated progeny of Ca 300

PROVINCE Plankton production and species distribution in the limnological p 379
PROVINCE The vegetation regions of the prairie province. Bot. Gaz. 25: 381- 585

PSORALEA A study of the genus Psoralea in Oklahoma. M. S. Thesis, Univ. Okl 270

PTERIDIOPHYTES Bryophytes and pteridiophytes of some west central Oklahoma 405

PTEROGONUM A taxonomic study of the subgenus Pterogonum of the genus Eriogo 311

PYRRHOPAPPUS Comparative pollination ecology of Pyrrhopappus geiseri and Py 38
PYRRHOPAPPUS The host plant-insect ecology of Pyrrhopappus (Asteraceae: Lac 861
PYRRHOPAPPUS Adaptation of the life-history of Schinia mitis to its host pl 862

PYRRHOPHYTA A survey of auxotrophy in five freshwater dinoflagellates (Pyrr 322

QUADRAT An evaluation of the quadrat method in the blackjack-post oak fores 152
QUADRAT Efficiencies of the point-center quarter and quadrat methods in for 446
QUADRAT A statistical method for determining quadrat size and adequacy of s 606

QUANTITATIVE A quantitative comparison of the soil algae of central Oklahom 213
QUANTITATIVE A quantitative comparison between an oak forest and an oak sav 352
QUANTITATIVE The quantitative phenology of two plant communities in Osage C 355
QUANTITATIVE A quantitative analysis of bottomland communities in south-cen 567

QUERCUS Vegetation response following winged elm (Ulmus alata Michx.) and o 172
QUERCUS A comparative study of the water relations of Quercus marilandica a 283
QUERCUS Geographic distribution of two species of oaks in Oklahoma in relat 284
QUERCUS Biomass, annual net primary production, and dynamics of six mineral 351
QUERCUS A quantitative comparison between an oak forest and an oak savannah 352
QUERCUS The significance of vegetative reproduction in Quercus. Madroño 11: 490
QUERCUS Oak (Quercus spp.) sprout growth rates on a central Oklahoma shallo 586
QUERCUS An evaluation of the variable-radius and paired-tree methods in the 619
QUERCUS Some variations of the fruit of Quercus macrocarpa. M. S. Thesis, O 708
QUERCUS Preliminary ecological study of the shinnery oak area of western Ok 842

RAGWEED Weather patterns and ragweed production in Tulsa, Oklahoma. Ann. Al 87
RAGWEED Ragweed pollen and weather patterns in Tulsa, Oklahoma. Bull. Ecol. 88
RAGWEED An assessment of weather patterns and ragweed pollen in Tulsa over 89

RAILROAD Report of Capt. R. B. Marcy on the railroad from Fort Smith to San 442
RAILROAD [reports of railroad route expeditions]. 33rd. Cong., 2nd.Sess., S 787
RAILROAD Reports of explorations and surveys ... [long title – deals with 788
RAILROAD [long title concerning exploration of railroad routes]. 33rd. Cong 789

RAINFALL Correlation analysis of rainfall and annual ring index of central 350

RANCH Conservation ranching in the Oklahoma panhandle. J. Range Manage. 3: 31
RANCH An annotated plant list for Adams Ranch. M. S. Thesis, Ft. Hays Kansa 198
RANCH A vegetational analysis of the Phillips Agricultural Demonstration Pr 248

RANGE Distribution and control of several woody plants in Texas and Oklahom 25
RANGE Vegetation-soil relationships in Flint Hills bluestem pastures. J. Ra 27
RANGE Pasture and range plants. Phillips Petrol. Co., Bartlesville, OK. {pr 29
RANGE Conservation ranching in the Oklahoma panhandle. J. Range Manage. 3: 31
RANGE The range of Cypripedium kentuckiense. Amer. Orchid Soc. Bull. 54(10) 32
RANGE Evaluation of sampling techniques on tall-grass prairie. J. Range Man 48
RANGE Vegetation analysis of bobwhite habitat of northeastern Oklahoma rang 138
RANGE Canopy structure of a tall-grass prairie. J. Range Manage. 27: 313-31 147
RANGE Vegetational responses following winged elm and oak control in Oklaho 173
RANGE Manipulating central Oklahoma rangeland for bobwhite quail. M. S. The 185
RANGE Influence of grazing on plant succession of rangelands. Bot. Rev. 26: 204
RANGE New methods of brush control for more grass. J. Range Manage. 3: 46-5 210
RANGE Interseeding and paraquat effects on central and eastern rangeland ve 224
RANGE The mesquite problem in the southwest. J. Range Manage. 3: 60-71. {Pr 234
RANGE The effects of one year's nitrogen fertilization on native vegetation 251
RANGE Burning and fertilization for range improvement in central Oklahoma. 271
RANGE Water stress of tallgrass prairie plants in central Oklahoma. J. Rang 277
RANGE Production characteristics of Oklahoma forages: Native range. Okla. A 291
RANGE Forage production, vegetative composition, and plant vigor in relatio 304
RANGE The claypan range site in northern Osage County, Oklahoma. J. Range M 305
RANGE Effect of grazing intensity on plant composition, vigor, and producti 306
RANGE Retrogression of native range vegetation in the sandy high plains. M. 317
RANGE Effect of fertilization on native grass pastures in Oklahoma. J. Rang 331
RANGE Developmental growth of four species of range grasses in north-centra 334
RANGE Grazing use checks on the Wichitas. J. Range Manage. 16: 142-145. {SW 378
RANGE Energy analysis of Oklahoma rangelands and improved pastures. J. Rang 384
RANGE Populations and range effects of rodents on the sand sagebrush grassl 457
RANGE A summary of fire and forage research on shinnery oak rangelands. Pro 468
RANGE Spraying 2,4-D by airplane on sand sagebrush and other plants of the 469

RANGE Fourteen year summary of range improvement studies. Bull. U. S. South 470
RANGE Range and pasture studies and grazing recommendations for the souther 474
RANGE Germination of downy brome from southern Kansas, central Oklahoma, an 483
RANGE Range deterioration in south central Oklahoma. M. S. Thesis, Univ. Ok 494
RANGE Some experiments and observations on revegetation of Oklahoma range l 515
RANGE The relation of grazing to plant succession in the tall grass prairie 551
RANGE Effects of fencing and plowing on plant succession in a revegetating 557
RANGE Oak (Quercus spp.) sprout growth rates on a central Oklahoma shallow 586
RANGE Growth and development of five species of range grasses in central Ok 598
RANGE Plant community structure. In: N. R. French (ed) Preliminary analysis 627
RANGE Pattern of retrogression of native vegetation in north central Oklaho 688

RARE Notes on rare Texas and Oklahoma plants. Sida 9: 76-86. {flora} 436
RARE Coronastrum: a rare find. Trans. Amer. Micros. Soc. 96: 163. {aquatic 568
RARE Herbaceous plants. In: Rare and endangered vertebrates and plants of O 700
RARE Gloeodendron: A rare alga in Oklahoma. Proc. Okla. Acad. Sci. 57: 169- 855
RARE Endangered and threatened plants of Oklahoma. Dept. Botany-Micro., Uni 859

RATTLEBOX Toxic plants of Oklahoma: Groundsels and rattlebox. J. Okla. Vet. 104

RAVINE Fluctuations in biotic communities. III. Aspection in a ravine sere 124

RECLAMATION Investigations in erosion control and reclamation of eroded lan 174
RECLAMATION Seed hay for regrassing eroded land. Proc. Okla. Acad. Sci. 25: 206
RECLAMATION The natural revegetation of lead/zinc mine spoil in northeaster 255
RECLAMATION Vegetation resources including endangered and threatened specie 312
RECLAMATION Field investigations, McGee Creek Project, Oklahoma, Calamovilf 313
RECLAMATION An environmental study of the W. C. Austin Irrigation Project o 314
RECLAMATION Revegetation of coal stripped land near Henryetta, Oklahoma. Pr 439
RECLAMATION Revegetation of eroded lands in the High Plains of Texas and Ok 508
RECLAMATION Revegetation of strip mine spoils. M. S. Thesis, Okla. State Un 702

REDBED A floristic study of the vascular plants of the Gypsum Hills and Red 35
REDBED Floristic components of the Gypsum Hills and Redbed Plains of southw 36

REDBUD Genecology of eastern redbud (Cercis canadensis). Ecology 63: 962-97 190

REDUCTASE Differences in nitrate reductase activity between species of diff 699

REESTABLISHING Methods of reestablishing buffalo grass on cultivated land i 647

REFUGE Vegetation of the Salt Plains Wildlife Refuge, Jet, Oklahoma. Ph. D. 33
REFUGE Relationships of the woody vegetation of the Wichita Mountains Wildl 84
REFUGE Relationships of the woody vegetation of the Wichita Mountains Wildl 85
REFUGE Some additions to Eskew's list of flowering plants of the Wichita Mo 90
REFUGE The effect of salinity on the distribution of algae at the Salt Plai 113
REFUGE Grassland communities of the Wichita Mountains Wildlife Refuge. Ph. 162
REFUGE Influence of soils and parent materials on grasslands of the Wichita 163
REFUGE The vegetation of the Wichita Mountains Wildlife Refuge. M. S. Thesi 186
REFUGE The flowering plants of the Wichita Mountains Wildlife Refuge. Amer. 217
REFUGE The occurrence of persimmon and buckeye in the Wichita Mountains Wil 285
REFUGE Influence of a wildfire in the Wichita Mountains Wildlife Refuge. Ec 554
REFUGE National Wildlife Refuges – What future? Ecosystems 6: 4-11. {} 727

REGENERATION Natural regeneration of loblolly pine 230 miles northwest of i 584

REGIMENT Journal of the expedition of the regiment of dragoons into Indian 839

REGION Ecological relationships of the Bird Creek region, northeastern Okla 54

REGION Revegetation and gully-control experiments in the Red Plains Region 209
REGION Relation of climatic conditions, soil characteristics, and tree deve 294
REGION Natural regions of the United States and Canada. W. H. Freeman Co., 333
REGION The development of aquatic ecoregions in Oklahoma. Ph. D. Thesis, Un 342
REGION The native orchids of the prairies and plains region of North Americ 430
REGION The vegetation regions of the prairie province. Bot. Gaz. 25: 381-39 585
REGION The vegetation of the Mesa de Maya region of Colorado, New Mexico, a 639
REGION Some botanical studies in the Black Mesa region of Oklahoma. Rhodora 640
REGION The natural vegetation of the Great Plains region. Ann. Assoc. Amer. 668
REGION Biogeographic regions of Texas and Oklahoma. Ecology 31: 427-433. {a 833
REGION Oklahoma climagraphs and biotic regions. Proc. Okla. Acad. Sci. 5: 9 835

REGRASSING Seed hay for regrassing eroded land. Proc. Okla. Acad. Sci. 25: 206
REGRASSING Data and observations concerning regrassing abandoned cultivated 475
REGRASSING Progress of regrassing in the Southern Great Plains. Soil Cons. 829

RELICT Effect of mowing on a relict tall grass prairie. Proc. Okla. Acad. S 164
RELICT Relationships of sugar maples (Acer saccharum and A. grandidentatum) 252
RELICT The microclimate of a relict stand of sugar maple in Devil's Canyon 600

REPLACEMENT Replacement of a population of Johnson grass by a vine-forb com 556

REPRODUCTION Reproduction of woody plants in the absence of grazing. Proc. 119
REPRODUCTION Vegetative habit and reproduction of Desmidium grevillii (Kütz 154
REPRODUCTION The significance of vegetative reproduction in Quercus. Madroñ 490

REPRODUCTIVE Reproductive studies in the genus Callirhoe (Malvaceae). Ph. D 26
REPRODUCTIVE Toxic plants of Oklahoma: Species causing reproductive disorde 109
REPRODUCTIVE Reproductive biology of Commelina erecta (Commelinaceae). pp. 449
REPRODUCTIVE The reproductive biology of Proboscidea louisianica (Martyniac 573
REPRODUCTIVE The reproductive biology of Proboscidea louisianica (Martyniac 574
REPRODUCTIVE Cytogeography and reproductive biology of Achillea millefolium 577
REPRODUCTIVE Distribution, habitat, and reproductive biology of Phlox oklah 703

RESPIRATORY Toxic plants of Oklahoma: Species causing acute respiratory dis 110

RESTORATION Restoration of eroded prairie with digested sludge. Proc. Okla. 377

RETROGRESSION Retrogression of native range vegetation in the sandy high pl 317
RETROGRESSION The pattern of retrogression of native vegetation in north ce 686
RETROGRESSION Pattern of retrogression of native vegetation in north centra 688

REVEGETATING Effects of fencing and plowing on plant succession in a revege 557
REVEGETATING Effects of oat straw and native hay mulches on plant compositi 610

REVEGETATION Revegetation of abandoned fields in Kansas and Oklahoma. Amer. 67
REVEGETATION Investigations in erosion control and reclamation of eroded la 174
REVEGETATION Seed hay for regrassing eroded land. Proc. Okla. Acad. Sci. 25 206
REVEGETATION Revegetation and gully-control experiments in the Red Plains 209
REVEGETATION The natural revegetation of lead/zinc mine spoil in northeaste 255
REVEGETATION Revegetation of unreclaimed coal strip mines in Oklahoma. II. 256
REVEGETATION A comparison of forest vegetation on orphan strip mines and ad 299
REVEGETATION Revegetation of unreclaimed coal strip-mines in Oklahoma. I. V 348
REVEGETATION Revegetation of coal stripped land near Henryetta, Oklahoma. P 439
REVEGETATION Revegetation of eroded lands in the High Plains of Texas and O 508
REVEGETATION Some experiments and observations on revegetation of Oklahoma 515
REVEGETATION Soil factors preventing revegetation of a denuded area near an 532
REVEGETATION Natural revegetation of abandoned farmland in the central and 650

REVEGETATION Revegetation of strip mine spoils. M. S. Thesis, Okla. State U 702
REVEGETATION A study of revegetation in the floodplain of the South Canadia 807

RHODOPHYTA Lemnaea fuscina Borg (Rhodophyta) in Oklahoma. Proc. Okla. Ac 730

RHUS The effect of fire on the growth of Rhus glabra L. and Smilax bona-nox 332
RHUS A study of the role of Rhus copallina in the community dynamics of the 565
RHUS The role of Rhus copallina in the dynamics of the forest-prairie ecoto 566

ROADSIDE Roadside flowers of Oklahoma. C & J Printing Co., Lawton, OK. {flo 450
ROADSIDE Roadside flowers of Oklahoma. Vol. 2. C & J Printing Co., Lawton, 451
ROADSIDE Roadside wild fruits of Oklahoma. Univ. Okla. Press, Norman. {flor 452
ROADSIDE Roadside trees and shrubs of Oklahoma. Univ. Okla. Press, Norman. 453
ROADSIDE Wildlife and roadside erosion in central Oklahoma. Proc. Okla. Aca 694

ROCKY Account of an expedition from Pittsburg to the Rocky Mountains, perfo 340
ROCKY Grass stand establishment on rough, rocky, wooded lands in eastern Ok 423
ROCKY Some account of a collection of plants made during a journey to and f 764

ROOTING Natural variation in rooting ability of western provenances of shor 223

RUBIACEAE Galium pedemontanum (Bell.) All. (Rubiaceae) reported from Oklaho 395
RUBIACEAE The genus Galium (Rubiaceae) in the southeastern United States. P 396
RUBIACEAE Hedyotis ouachitana (Rubiaceae): a new species from the Ouachita 698

SAGE Rodent populations and their reactions to grazing intensities on sand 242
SAGE The mammals of Harmon County, Oklahoma. Proc. Okla. Acad. Sci. 49: 42- 444
SAGE Populations and range effects of rodents on the sand sagebrush grassla 457
SAGE Progress in shinnery oak and sand sage control at Woodward. Prog. Rept 467
SAGE Spraying 2,4-D by airplane on sand sagebrush and other plants of the s 469
SAGE Annotated checklist of the vascular plants of Little Sahara State Park 679
SAGE A comparison of the predominant soil microflora in eight vegetative co 706
SAGE Identification and mapping of the extant flora at the Deer Creek archa 776

SAGEBRUSH Rodent populations and their reactions to grazing intensities on 242
SAGEBRUSH Populations and range effects of rodents on the sand sagebrush gr 457
SAGEBRUSH Spraying 2,4-D by airplane on sand sagebrush and other plants of 469

SAHARA Annotated checklist of the vascular plants of Little Sahara State Pa 679

SALINITY The effect of salinity on the distribution of algae at the Salt Pl 113
SALINITY Salinity tolerance of marsh plants in Kansas and Oklahoma. Ecology 791

SALIX An autecological study of Salix nigra, especially factors that might 471
SALIX Factors limiting the distribution of Salix nigra. Bull. Torrey Bot. C 472
SALIX Vegetational changes in a black willow forest over a four year period 553

SALT Vegetation of the Salt Plains Wildlife Refuge, Jet, Oklahoma. Ph. D. D 33
SALT The effect of salinity on the distribution of algae at the Salt Plains 113
SALT Diatoms, including salt-water taxa, from southwestern Oklahoma. Proc. 385
SALT The ecology of the western Oklahoma salt plains. Pub. Univ. Okla. Biol 519
SALT Salinity tolerance of marsh plants in Kansas and Oklahoma. Ecology 47: 791
SALT Species-soil relationships in the Great Salt Plains of northern Oklaho 792

SALT-WATER Diatoms, including salt-water taxa, from southwestern Oklahoma. 385

SAND Rodent populations and their reactions to grazing intensities on sand 242
SAND Differential selection in juniper populations from the Baum Limestone 282
SAND Retrogression of native range vegetation in the sandy high plains. M. 317
SAND The mammals of Harmon County, Oklahoma. Proc. Okla. Acad. Sci. 49: 42- 444

SAND Populations and range effects of rodents on the sand sagebrush grassla 457
SAND Progress in shinnery oak and sand sage control at Woodward. Prog. Rept 467
SAND Spraying 2,4-D by airplane on sand sagebrush and other plants of the s 469
SAND A new sandbur from western Oklahoma: Cenchrus albertsonii. Amer. J. Bo 644
SAND Vegetation of the Woods County, Oklahoma sand dunes. Ph. D. Thesis, Un 677
SAND Annotated checklist of the vascular plants of Little Sahara State Park 679
SAND A comparison of the predominant soil microflora in eight vegetative co 706
SAND A comparison of algal succession between farm ponds with sandstone and 856
SAND A comparison of algal composition and seasonal succession between pond 857

SANDBUR A new sandbur from western Oklahoma: Cenchrus albertsonii. Amer. J. 644

SANDSTONE A comparison of algal succession between farm ponds with sandston 856
SANDSTONE A comparison of algal composition and seasonal succession between 857

SANDY Retrogression of native range vegetation in the sandy high plains. M. 317

SANS BOIS Vascular plants of southeastern Oklahoma from the Sans Bois to th 478

SANTA FE Commerce of the prairies: or, The journal of a Santa Fe trader, du 274
SANTA FE Report of Capt. R. B. Marcy on the railroad from Fort Smith to San 442

SARCOSTEMMA A new species of Sarcostemma from Oklahoma. Rhodora 51: 58- 814

SAVANNA Distribution and control of several woody plants in Texas and Oklah 25
SAVANNA Seasonal aspects in certain prairie and savannah communities. M. S. 121
SAVANNA The savannah concept and its use. Ecology 38: 435-442. {classificat 200
SAVANNA A floristic study of the woody vegetation of the North American Cro 302
SAVANNA A quantitative comparison between an oak forest and an oak savannah 352
SAVANNA Prairie dogs in shinnery (oak scrub) savannah. Ecology 23: 110-115. 522
SAVANNA The savanna concept in Oklahoma. Ecology 43: 774-775. {all classifi 548
SAVANNA Influence of a wildfire in the Wichita Mountains Wildlife Refuge. E 554
SAVANNA Oak (Quercus spp.) sprout growth rates on a central Oklahoma shallo 586

SCHIZACHYRIUM Analysis of morphological variation in Schizachyrium scopariu 77
SCHIZACHYRIUM The influence of burning on tillering and aboveground biomass 717

SCRUB Prairie dogs in shinnery (oak scrub) savannah. Ecology 23: 110-115. { 522

SEASON Seasonal aspects in certain prairie and savannah communities. M. S. 121
SEASON Insect distribution and seasonal succession in overgrazed and normal 157
SEASON Geographic distribution of two species of oaks in Oklahoma in relati 284
SEASON A preliminary report on the seasonal aspects of six habitats near No 308
SEASON Habitat used by lesser prairie chickens for feeding related to seaso 357
SEASON Wind patterns and the seasonal occurrence of Juniperus pollen in Tul 402
SEASON Seasonal algal flora on the campus of Catholic College, Guthrie, Okl 440
SEASON The phenology of woody plants in Norman, Oklahoma during the spring 544
SEASON Distribution, seasonal abundance, and hosts of the Gulf Coast tick i 662
SEASON The ecology and behavior of the lone star tick. (Acarina: Ixodidae). 663
SEASON Establishment of warm and cool season grass pastures on wooded sites 664
SEASON The structure and function of ten western North American grasslands. 691
SEASON A comparison of algal composition and seasonal succession between po 857

SEDIMENT Sediment-precipitation-runoff relationships for the three ephemera 638
SEDIMENT Plant development and sequence in artificial ponds with and withou 769

SEED Seed predation, seed dispersal, and disturbance in grasslands: a comme 143
SEED After-ripening and germination of seed of Juniperus virginiana L. and 160
SEED Toxic plants of Oklahoma: Cotton and cottonseed. J. Okla. Vet. Med. As 202

SEED Seed hay for regrassing eroded land. Proc. Okla. Acad. Sci. 25: 43-45. 206
SEED Interseeding and paraquat effects on central and eastern rangeland veg 224
SEED Germination studies of "Caddo" sugar maple (Acer saccharum) seeds. See 482
SEED The seeding, development, and utilization of native grasses in Kay Cou 485
SEED Germination of weed seeds with reference to temperatures and storage c 493
SEED A study of viable seeds in a selection of birds' nests. Ph. D. Thesis, 507
SEED Some experiments and observations on revegetation of Oklahoma range la 515
SEED Seed dispersal and mineral nutrition in succession in abandoned fields 611
SEED Mechanism of seed dormancy in Ambrosia artemisiifolia. Ph. D. Diss., U 844

SENNA Toxic plants of Oklahoma: Coffeeweeds and sennas. J. Okla. Vet. Med. 102

SERE Fluctuations in biotic communities. III. Aspection in a ravine sere in 124

SHELTERBELTS Honeylocust (Gleditsia triacanthos L.) in shelterbelts of west 243

SHINNERY The mammals of Harmon County, Oklahoma. Proc. Okla. Acad. Sci. 49: 444
SHINNERY Shinnery oak can be controlled. Proc. Southern Weed Conf. 9: 95-98 466
SHINNERY Progress in shinnery oak and sand sage control at Woodward. Prog. 467
SHINNERY A summary of fire and forage research on shinnery oak rangelands. 468
SHINNERY The significance of vegetative reproduction in Quercus. Madroño 11 490
SHINNERY Prairie dogs in shinnery (oak scrub) savannah. Ecology 23: 110-115 522
SHINNERY A comparison of the predominant soil microflora in eight vegetativ 706
SHINNERY Preliminary ecological study of the shinnery oak area of western O 842
SHINNERY A preliminary study of the shinnery in Oklahoma. Southw. Natur. 5: 843

SHORTGRASS The grassland biome. Ecol. Monogr. 10: 617-684. {prairie shortgr 127
SHORTGRASS Atlas of the flora of the Great Plains. Iowa State Univ. Press, 464
SHORTGRASS Flora of the Great Plains. Univ. Kansas Press, Lawrence. {flora 465
SHORTGRASS Annotated checklist of the vascular plants of Little Sahara Stat 679

SHORTLEAF Introgressive hybridization between shortleaf and loblolly pine i 1
SHORTLEAF Herbicides for release of shortleaf pine and native grass. Weeds 207
SHORTLEAF Bio-economic analysis of even aged shortleaf pine stands in south 211
SHORTLEAF Natural variation in rooting ability of western provenances of sh 223
SHORTLEAF Phenotypic variation and relationships of several variables in sh 458
SHORTLEAF Relationship of soil properties to site index of shortleaf pine a 526
SHORTLEAF Shortleaf pine (Pinus echinata) in Bryan County, Oklahoma. Southw 728
SHORTLEAF Shortleaf pine site index – Soils and plant associations on the 851

SHRUB Relationships of the woody vegetation of the Wichita Mountains Wildli 85
SHRUB Toxic plants of Oklahoma: Ornamental shrubs. J. Okla. Vet. Med. Assoc 101
SHRUB Reproduction of woody plants in the absence of grazing. Proc. Okla. A 119
SHRUB Notes on trees, shrubs, and vines in the Cherokee Nation. Trans. Kans 266
SHRUB A nature study outline on trees and shrubs for the first six grades. 290
SHRUB An analysis of selected Oklahoma upland forest stands including both 375
SHRUB Trees and shrubs of the Southeastern Oklahoma Protective Unit. Proc. 416
SHRUB Roadside trees and shrubs of Oklahoma. Univ. Okla. Press, Norman. {fl 453
SHRUB Competitive and allelopathic suppression of understory by Oklahoma oa 476
SHRUB Bottomland forests of north-central Oklahoma. Ecology 46: 708-714. {N 605
SHRUB Phytosociological analysis of Oklahoma upland forest species. Ecology 633
SHRUB Shrub distribution patterns in Oklahoma with comparisons to those exh 657
SHRUB The trees and shrubs of Oklahoma. Okla. Geol. Surv. Circ. 4: 7-41. {f 666
SHRUB The woody plants of Oklahoma. Ph. D. Thesis, Univ. Calif., Berkeley. 683
SHRUB Atlas of the woody plants of Oklahoma. Okla. Biol. Surv., Norman. {tr 845

SIDA Sida in Oklahoma. Rhodora 50: 99. {flora} 136

SILT Silting and forest succession on Deep Fork in southwestern Creek Count 229
SILT Effect of silting on tree development in the flood plain of the North 293
SILT Effect of silt on natural vegetation and drainage in the flood plain o 297

SILVERLEAF Silverleaf nightshade (Solanum eleagnifolium) origin, distributi 71

SILVICS Silvics of forest trees of the United States. Agr. Handbook No. 271 241

SLOPE Vegetation in relation to slope exposure and geology in the Arbuckle 335

SLUDGE Restoration of eroded prairie with digested sludge. Proc. Okla. Acad 377

SMELTER Plant succession following destruction of vegetation by toxins from 827

SMILAX The effect of fire on the growth of Rhus glabra L. and Smilax bona-n 332

SNEEZEWEED Toxic plants of Oklahoma: Bitterweeds and sneezeweeds. J. Okla. 98

SOIL Vegetation-soil relationships in Flint Hills bluestem pastures. J. Ran 27
SOIL Standing state and cycling of nitrogen in soil-vegetation components o 61
SOIL The relation of plant communities to soil erosion in Cleveland County, 65
SOIL Algae as pioneers in plant succession and their importance in erosion 66
SOIL Relationships of the woody vegetation of the Wichita Mountains Wildlif 84
SOIL Relationships of the woody vegetation of the Wichita Mountains Wildlif 85
SOIL Micronutrients in soils of the tall-grass prairie of Oklahoma. Proc. O 120
SOIL A soil improvement program for Oklahoma. Okla. A & M Coll. Circ. No. 4 131
SOIL The effect of moisture on the experimental determination of the metabo 158
SOIL Influence of soils and parent materials on grasslands of the Wichita M 163
SOIL Investigations in erosion control and reclamation of eroded land at th 174
SOIL A comparison of post oak-blackjack oak communities in two major soil t 199
SOIL Seed hay for regrassing eroded land. Proc. Okla. Acad. Sci. 25: 43-45. 206
SOIL Revegetation and gully-control experiments in the Red Plains Region. 209
SOIL A comparison of the soil fungi of a tall-grass prairie and of an aband 212
SOIL A quantitative comparison of the soil algae of central Oklahoma prairi 213
SOIL Soils of Oklahoma. Okla. State Univ. Misc. Pub. 56, Stillwater. {soil 272
SOIL Benchmark and key soils of Oklahoma. Okla. State Univ. Agr. Exp. Sta. 273
SOIL A comparison of soil microfungi of two grassland areas in central Okla 288
SOIL Relation of climatic conditions, soil characteristics, and tree develo 294
SOIL Effects of chemical composition of soil on micronutrients and other el 298
SOIL Some vegetation-environment relationships in the upland forests of Okl 344
SOIL Biomass, annual net primary production and dynamics of six mineral ele 345
SOIL Revegetation of unreclaimed coal strip-mines in Oklahoma. I. Vegetatio 348
SOIL Acetylene reduction (nitrogen fixation) and old-field succession in ce 359
SOIL Acetylene reduction (N-fixation) of glucose amended soils from central 361
SOIL Acetylene reduction (N-fixation) in soil and old-field succession in c 362
SOIL Symbiotic and asymbiotic N2-fixation in a tall grass prairie. Soil Bio 363
SOIL Vegetation and soil conditions of prairie and pasture plots in central 367
SOIL A comparison of organic carbon, hydrogen ion concentration, and volume 368
SOIL Restoration of eroded prairie with digested sludge. Proc. Okla. Acad. 377
SOIL Relation between soil fungi and seed plants in three successional fore 437
SOIL Relation between soil fungi and seed plants in three successional fore 438
SOIL A comparison of the soil fungi of a virgin prairie and an abandoned fi 461
SOIL An investigation of the soil microflora of two grassland plots. Proc. 462
SOIL Condensed tannins in soil: inputs and effects on microbial populations 486
SOIL The Gulf Coastal Plain Ultisol and Alfisol ecosystem: surface geology, 502
SOIL A study of soil compaction in a tall grass prairie. M. S. Thesis, Univ 506

SOIL Revegetation of eroded lands in the High Plains of Texas and Oklahoma. 508
SOIL Relationship of soil properties to site index of shortleaf pine and di 526
SOIL Soil enzymes in relation to old field succession: amylase, cellulase, 531
SOIL Soil factors preventing revegetation of a denuded area near an abandon 532
SOIL Edaphic and topographic effects on forest communities in Payne County, 536
SOIL The relationship of plant species heterogeneity to soil variation in b 580
SOIL Correlation between vegetation and geologic formations in Oklahoma. Ok 597
SOIL Effects of oat straw and native hay mulches on plant composition and e 610
SOIL Causes of decreases in residual carbohydrase activity in soil during o 612
SOIL Ecosystem analysis of the tallgrass prairie: nitrogen cycle. Ecology 6 636
SOIL Sediment-precipitation-runoff relationships for the three ephemeral fo 638
SOIL Calcium, magnesium, phosphorus, and potassium budgets of a tall grass 674
SOIL A model for phosphorus and potassium flux in a tall-grass prairie. Sou 675
SOIL Wildlife and roadside erosion in central Oklahoma. Proc. Okla. Acad. S 694
SOIL Biotic and physiographic succession on abandoned eroded farmland. Ecol 697
SOIL A comparison of the predominant soil microflora in eight vegetative co 706
SOIL A continuation of the investigation of the soil microflora of two gras 718
SOIL The relation of forest vegetation to soils and geology in the gulf coa 729
SOIL Fungi of various depths in typical Cleveland County, Oklahoma soils. P 763
SOIL Microbial decomposition of ferulic acid in soil. J. Chem. Ecol. 1: 41- 771
SOIL Species-soil relationships in the Great Salt Plains of northern Oklaho 792
SOIL Soils and men. Yearbook of agriculture. U. S. Govt. Printing Office, W 793
SOIL Primary succession on granite outcrops in southwestern Oklahoma. Bull. 797
SOIL Progress of regrassing in the Southern Great Plains. Soil Cons. 5: 57- 829
SOIL A survey of the soil algae of two vegetation types in central Oklahoma 847
SOIL An exploratory study on soil algae. Ecology 38: 309-313.`{SC grassland 848
SOIL Shortleaf pine site index – Soils and plant associations on the coast 851

SOIL-VEGETATION Standing state and cycling of nitrogen in soil-vegetation c 61

SOLANACEAE The pollination ecology of Solanum rostratum (Solanaceae). Amer. 69
SOLANACEAE Systematic studies in Physalis (Solanaceae). Ph. D. Thesis, Univ 716

SOLANUM Ecological studies of plant inhibition by Solanum rostratum. Ph. D. 15
SOLANUM The pollination ecology of Solanum rostratum (Solanaceae). Amer. J. 69
SOLANUM Silverleaf nightshade (Solanum eleagnifolium) origin, distribution, 71

SOLIDAGO New species, new combinations, and notes on the goldenrods (Eutham 725
SOLIDAGO Solidago (Asteraceae) in Oklahoma and Texas. Sida 10: 223-251. {fl 726
SOLIDAGO Solidago flaccidifolia Small (Asteraceae) new to Oklahoma and othe 798

SORGHASTRUM Carbon dioxide exchange characteristics of some prairie grass s 630

SORGHUM Replacement of a population of Johnson grass by a vine-forb communi 556
SORGHUM Possible mechanisms of the exclusion of Johnson grass by tall grass 660

SPECIES Nutrient and biomass allocation strategies for five grass species i 8
SPECIES Differential response of woody and herbaceous species to summer and 9
SPECIES Nutrient and biomass allocation in five grass species in an Oklahom 11
SPECIES Comparative water relations of four species of upland deciduous tre 23
SPECIES The influence of varying light intensities on the vegetative growth 49
SPECIES A list of some of the most interesting species of plants collected 114
SPECIES Growth of ten species of preferred deer browse plants under various 132
SPECIES Relationships of two isolated groups of sugar maple in central Okla 183
SPECIES Relationships of two isolated groups of sugar maples (Acer saccharu 184
SPECIES A new species of grama grass. Bot. Gaz. 91: 103-105. {Comanche SW f 225

SPECIES A new species of Streptanthus. Rhodora 58: 354-355. {flora} 262
SPECIES Geographic distribution of two species of oaks in Oklahoma in relat 284
SPECIES Electrophoretic analysis of soluble enzymes in five freshwater dino 303
SPECIES Vegetation resources including endangered and threatened species of 312
SPECIES Developmental growth of four species of range grasses in north-cent 334
SPECIES Plankton production and species distribution in the limnological pr 379
SPECIES Net production relations of two tree species in southern Oklahoma. 424
SPECIES Growth and development of five species of range grasses in central 598
SPECIES A report on the threatened and endangered plant species in the BLM 631
SPECIES Drought survival of native grass species in the central and souther 648
SPECIES The prairie fringed orchids: A pollinator-isolated species pair. Rh 681
SPECIES Some effects of differential clipping on six native grasses and on 692
SPECIES Hedyotis ouachitana (Rubiaceae): a new species from the Ouachita Mo 698
SPECIES New species, new combinations, and notes on the goldenrods (Euthami 725
SPECIES Veratrum woodii (Melanthium woodii). Status report. Endangered Spec 742
SPECIES Amorpha ouachitensis. Status report. Endangered Species Office, U. 743
SPECIES Hypoxis longii. Status report. Endangered Species Office, U. S. Fis 744
SPECIES Calamovilfa arcuata. Status report. Endangered Species Office, U. S 745
SPECIES Monarda stipitatoglandulosa. Status report. Endangered Species Offi 746
SPECIES Phlox oklahomensis. Status report. Endangered Species Office, U. S. 747
SPECIES Lesquerella angustifolia. Status report. Endangered Species Office, 748
SPECIES Cuscuta attenuata. Status report. Endangered Species Office, U. S. 749
SPECIES Carex latebracteata. Status report. Endangered Species Office, U. S 750
SPECIES Streptanthus squamiformis. Status report. Endangered Species Office 751
SPECIES Phlox pilosa var. longipilosa. Status report. Endangered Species Of 752
SPECIES Field surveys and status evaluation on proposed and candidate endan 781
SPECIES A new species of Euphorbia from Oklahoma. Rhodora 50: 63-64. {flora 812
SPECIES A new species of Sarcostemma from Oklahoma. Rhodora 51: 58-59. {flo 814
SPECIES A new species of Carex (Section Phyllostachyae) from Oklahoma. Rhod 821
SPECIES Monarda stipitatoglandulosa, a new species from Oklahoma. Rhodora 7 823

SPECIES-AREA The species-area curve. Ecology 36: 7-11. {methods phytosociol 614

SPECIES-SOIL Species-soil relationships in the Great Salt Plains of norther 792

SPERMATOPHYTA The spermatophyta of Custer County, Oklahoma. M. S. Thes 479
SPERMATOPHYTA The spermatophyta of Cherokee County, Oklahoma (exclusiv 802
SPERMATOPHYTA The spermatophyta of Oklahoma County, Oklahoma, exclusiv 810

SPHENOCLEA Additions to the Oklahoma flora: Sphenoclea zeylanica (Campanula 151

SPOIL Plant communities and the potential for ecotypic differentiation on l 254
SPOIL The natural revegetation of lead/zinc mine spoil in northeastern Okla 255
SPOIL Revegetation of strip mine spoils. M. S. Thesis, Okla. State Univ., S 702

SPOROBOLUS Allelopathic effects of Sporobolus pyramidatus on vegetational p 589
SPOROBOLUS Allelopathic effects of Sporobolus pyramidatus on vegetational p 590

SPRAYING Spraying 2,4-D by airplane on sand sagebrush and other plants of t 469
SPRAYING Establishment of warm and cool season grass pastures on wooded sit 664

SPRING The effect of early spring burning on vegetation in buffalo wallows. 142
SPRING Spring flora of central Oklahoma. Univ. Okla. Dup. Serv., Norman. {S 263
SPRING Notes on the phenology of middle western spring flowers. Proc. Okla. 268
SPRING Phenological notes for 1928 on middle western spring flowers. Proc. 269
SPRING Oklahoma spring flowers in autumn. Proc. Okla. Acad. Sci. 17: 60-62. 325
SPRING Simulated impact of management practices upon the tallgrass prairie. 540

SPRING The phenology of woody plants in Norman, Oklahoma during the spring 544
SPRING Phenology of spring weeds in Norman, Oklahoma. Southw. Natur. 1: 121 546

SPROUT Oak (Quercus spp.) sprout growth rates on a central Oklahoma shallow 586

STANDING Standing state and cycling of nitrogen in soil-vegetation componen 61

STREAM Spatial and temporal variation in productivity, species diversity, a 150
STREAM Variation with stream order in species composition, diversity, bioma 665

STREPTANTHUS A new species of Streptanthus. Rhodora 58: 354-355. {flora} 262
STREPTANTHUS Streptanthus squamiformis. Status report. Dept. of the Army, U 731
STREPTANTHUS Streptanthus squamiformis. Status report. Endangered Species O 751

STRESS Toxic plants of Oklahoma: Species causing acute respiratory distress 110
STRESS Water stress of tallgrass prairie plants in central Oklahoma. J. Ran 277

STRIP-MINES Revegetation of unreclaimed coal strip-mines in Oklahoma. I. Ve 348

STRIPPED Revegetation of coal stripped land near Henryetta, Oklahoma. Proc. 439

SUBSTRATES A comparison of algal succession between farm ponds with sandsto 856
SUBSTRATES A comparison of algal composition and seasonal succession betwee 857

SUCCESSION Plant inhibition by Johnson grass and its possible significance 2
SUCCESSION Plant inhibition by Johnson grass and its possible significance 3
SUCCESSION Inhibition of symbiotic nitrogen-fixation by gallic and gallotan 57
SUCCESSION Inhibition of symbiotic nitrogen fixation by gallic and tannic a 58
SUCCESSION The relation of plant communities to soil erosion in Cleveland C 65
SUCCESSION Algae as pioneers in plant succession and their importance in er 66
SUCCESSION Revegetation of abandoned fields in Kansas and Oklahoma. Amer. J 67
SUCCESSION Fluctuations in biotic communities. III. Aspection in a ravine s 124
SUCCESSION Succession in grasslands: thirty-two years of change in a centra 141
SUCCESSION Insect distribution and seasonal succession in overgrazed and no 157
SUCCESSION Influence of grazing on plant succession of rangelands. Bot. Rev 204
SUCCESSION Silting and forest succession on Deep Fork in southwestern Creek 229
SUCCESSION Plant communities and the potential for ecotypic differentiation 254
SUCCESSION A comparison of forest vegetation on orphan strip mines and adja 299
SUCCESSION Ecological studies on the Canadian River floodplain in Cleveland 309
SUCCESSION Mechanisms of secondary succession in a tall grass prairie. Ph. 336
SUCCESSION Vegetational changes in a black willow forest over a 23-year per 346
SUCCESSION Revegetation of unreclaimed coal strip-mines in Oklahoma. I. Veg 348
SUCCESSION Acetylene reduction (nitrogen fixation) and old-field succession 359
SUCCESSION Acetylene reduction (N-fixation) of glucose amended soils from c 361
SUCCESSION Acetylene reduction (N-fixation) in soil and old-field successio 362
SUCCESSION Changes in community structure in Oklahoma old field succession. 364
SUCCESSION The effects of periodic flooding on upland forest succession. M. 387
SUCCESSION Nitrogen fixation in selected prairie legumes as related to succ 419
SUCCESSION Relation between soil fungi and seed plants in three successiona 437
SUCCESSION Relation between soil fungi and seed plants in three successiona 438
SUCCESSION Revegetation of coal stripped land near Henryetta, Oklahoma. Pro 439
SUCCESSION Data and observations concerning regrassing abandoned cultivated 475
SUCCESSION Possible role of Ambrosia psilostachya on patterning and success 495
SUCCESSION Possible role of Ambrosia psilostachya on pattern and succession 496
SUCCESSION Relative effects of known plant inhibitors on species from first 514
SUCCESSION Soil enzymes in relation to old field succession: amylase, cellu 531
SUCCESSION Inhibitional effects of Digitaria sanguinalis and possible role 533

SUCCESSION Inhibitional effects of Digitaria sanguinalis and possible role 534
SUCCESSION Effects of certain plants of old-field succession on the growth 538
SUCCESSION The relation of grazing to plant succession in the tall grass pr 551
SUCCESSION Vegetational changes in a black willow forest over a four year p 553
SUCCESSION Replacement of a population of Johnson grass by a vine-forb comm 556
SUCCESSION Effects of fencing and plowing on plant succession in a revegeta 557
SUCCESSION Some aspects of structure and function in Oklahoma old-field suc 562
SUCCESSION Some aspects of structure and function in Oklahoma old-field suc 563
SUCCESSION A study of the role of Rhus copallina in the community dynamics 565
SUCCESSION The role of Rhus copallina in the dynamics of the forest-prairie 566
SUCCESSION Inhibition of nitrogen-fixing and nitrifying bacteria by seed pl 603
SUCCESSION Inhibition of nodulation of inoculated legumes by pioneer plant 607
SUCCESSION Some roles of allelopathic compounds in plant communities. Bioch 609
SUCCESSION Effects of oat straw and native hay mulches on plant composition 610
SUCCESSION Seed dispersal and mineral nutrition in succession in abandoned 611
SUCCESSION Causes of decreases in residual carbohydrase activity in soil du 612
SUCCESSION Plant succession and yield of living plant material in a plowed 618
SUCCESSION Plant succession on abandoned fields in central Oklahoma and in 643
SUCCESSION Methods of reestablishing buffalo grass on cultivated land in th 647
SUCCESSION Natural revegetation of abandoned farmland in the central and so 650
SUCCESSION Distribution of legumes as correlated with surface geology and p 673
SUCCESSION Biotic and physiographic succession on abandoned eroded farmland 695
SUCCESSION Biotic and physiographic succession on abandoned eroded farmland 697
SUCCESSION Differences in nitrate reductase activity between species of dif 699
SUCCESSION Primary succession on granite outcrops in southwestern Oklahoma. 797
SUCCESSION A study of revegetation in the floodplain of the South Canadian 807
SUCCESSION The vegetation of the lower levels of the South Canadian River i 808
SUCCESSION Plant succession following destruction of vegetation by toxins f 827
SUCCESSION Allelopathy as expressed by Helianthus annuus, and its role in o 849
SUCCESSION Allelopathy as expressed by Helianthus annuus and its role in ol 850
SUCCESSION A comparison of algal succession between farm ponds with sandsto 856
SUCCESSION A comparison of algal composition and seasonal succession betwee 857

SUGAR Relationships of two isolated groups of sugar maple in central Oklaho 183
SUGAR Relationships of two isolated groups of sugar maples (Acer saccharum 184
SUGAR Relationships of sugar maples (Acer saccharum and A. grandidentatum) 252
SUGAR Germination studies of "Caddo" sugar maple (Acer saccharum) seeds. Se 482
SUGAR The microclimate of a relict stand of sugar maple in Devil's Canyon i 600
SUGAR The microclimate of sugar maple stands in Oklahoma. Ecology 43: 19-25 601

SUNDEW A preliminary report on the ecology of a Drosera meadow. Proc. Okla. 39

SUNFLOWER Allelopathy as expressed by Helianthus annuus, and its role in ol 849
SUNFLOWER Allelopathy as expressed by Helianthus annuus and its role in old 850

SUPPRESSION Competitive and allelopathic suppression of understory by Oklah 476

SURFACE Petroleum and surface vegetation. Proc. Okla. Acad. Sci. 10: 110-11 267
SURFACE The Gulf Coastal Plain Ultisol and Alfisol ecosystem: surface geolo 502
SURFACE Model fish and wildlife plan for a surface coal mine in Oklahoma. O 655
SURFACE Distribution of legumes as correlated with surface geology and plan 673

SURVIVAL Drought survival of native grass species in the central and southe 648
SURVIVAL The ecology and behavior of the lone star tick (Acarina: Ixodidae) 661

SYMBIOTIC Inhibition of symbiotic nitrogen-fixation by gallic and gallotann 57

SYMBIOTIC Inhibition of symbiotic nitrogen fixation by gallic and tannic ac 58
SYMBIOTIC Symbiotic and asymbiotic N2-fixation in a tall grass prairie. Soi 363

TALL GRASS Tall grass defeats prairie dogs. Soil Cons. 20:103-105. {SW gras 22
TALL GRASS A survey of the tall grass prairies in northeastern Oklahoma. M. 83
TALL GRASS Effect of mowing on a relict tall grass prairie. Proc. Okla. Aca 164
TALL GRASS Effects of mowing on a native tall grass prairie in central Okla 247
TALL GRASS Mechanisms of secondary succession in a tall grass prairie. Ph. 336
TALL GRASS Symbiotic and asymbiotic N2-fixation in a tall grass prairie. So 363
TALL GRASS Effects of moderate grazing on the composition and plant product 369
TALL GRASS The effect of disturbances in tall grass prairie sites on an ind 503
TALL GRASS A study of soil compaction in a tall grass prairie. M. S. Thesis 506
TALL GRASS The relation of grazing to plant succession in the tall grass pr 551
TALL GRASS Causes of decreases in productivity in undisturbed tall grass pr 613
TALL GRASS A preliminary compartment model of a tall grass prairie: Osage 1 628
TALL GRASS Systems analysis of a tall grass prairie. Proc. Third Midw. Prai 629
TALL GRASS Herbage dynamics of a tall grass prairie: Osage, 1971. USIBP Gra 634
TALL GRASS Possible mechanisms of the exclusion of Johnson grass by tall gr 660
TALL GRASS Calcium, magnesium, phosphorus, and potassium budgets of a tall 674
TALL GRASS Effect of disturbance on microclimate and vegetation in tall gra 693

TALLGRASS Nutrient and biomass allocation strategies for five grass species 8
TALLGRASS Nutrient and biomass allocation in five grass species in an Oklah 11
TALLGRASS Evaluation of sampling techniques on tall-grass prairie. J. Range 48
TALLGRASS The tallgrass prairie orchids Platanthera leucophaea (Nutt.) Lind 70
TALLGRASS A partial inventory of insect populations in tallgrass prairie pa 81
TALLGRASS Interaction of disturbances in tallgrass prairie: a field experim 139
TALLGRASS Succession in grasslands: thirty-two years of change in a central 141
TALLGRASS Vegetation structure in a tallgrass prairie. M. S. Thesis, Univ. 146
TALLGRASS Water stress of tallgrass prairie plants in central Oklahoma. J. 277
TALLGRASS Atlas of the flora of the Great Plains. Iowa State Univ. Press, A 464
TALLGRASS Flora of the Great Plains. Univ. Kansas Press, Lawrence. {flora t 465
TALLGRASS A phenological comparison of mowed and unmowed tallgrass prairie 488
TALLGRASS Simulated impact of management practices upon the tallgrass prair 540
TALLGRASS Herbage dynamics of an Oklahoma tallgrass prairie: Osage 1972. US 635
TALLGRASS Ecosystem analysis of the tallgrass prairie: nitrogen cycle. Ecol 636

TANNIC Inhibition of symbiotic nitrogen-fixation by gallic and gallotannic 57
TANNIC Inhibition of symbiotic nitrogen fixation by gallic and tannic acid, 58

TANNINS Condensed tannins in soil: inputs and effects on microbial populati 486
TANNINS Inhibition of nitrification by climax ecosystems. II. Additional ev 616
TANNINS Inhibition of nitrification by climax ecosystems. III. Inhibitors o 617

TATSCHL A biogeographical boundary: The Tatschl Line. Trans. Kans. Acad. Sc 389

TAXODIUM Baldcypress (Taxodium distichum) in Oklahoma. Proc. Okla. Acad. Sc 413

TAXONOMIC Taxonomic studies in the Verbena stricta complex. Ph. D. Thesis, 37
TAXONOMIC A taxonomic investigation of the Polygonum pensylvanicum L. (Poly 201
TAXONOMIC A taxonomic study of the genus Juniperus in Oklahoma. M. S. Thesi 280
TAXONOMIC A taxonomic study of the subgenus Pterogonum of the genus Eriogon 311
TAXONOMIC A numerical taxonomic study of Monarda (Labiatae) in southern Okl 441
TAXONOMIC A taxonomic study of the Umbelliferae in Oklahoma. M. S. Thesis, 454
TAXONOMIC A taxonomic study of the vascular plants indigenous to, or natura 455

TAXONOMY Ecology and taxonomy of Water Canyon, Canadian County, Oklahoma 722

TECTONICS Distribution of the grasses (Gramineae) in relation to plate tect 520

TEMPERATURES Germination of weed seeds with reference to temperatures and s 493

TEMPORAL Spatial and temporal variation in productivity, species diversity, 150
TEMPORAL The ecology of Honey Creek: Temporal patterns of the travertine pe 596

TENSION Plant associations as site predictors in the pine-hardwood tension 187
TENSION Aspection in a central Oklahoma tension zone area. M. S. Thesis, Un 320

TEXOMA The composition of a black willow forest in south central Oklahoma. 547
TEXOMA Vegetational changes in a black willow forest over a four year perio 553

THORNAPPLES Toxic plants of Oklahoma: Thornapples and nightshades. J. Okla. 97

THREATENED Field study of the threatened and endangered plants of southeast 253
THREATENED Vegetation resources including endangered and threatened species 312
THREATENED A report on the threatened and endangered plant species in the B 631
THREATENED Endangered and threatened plants of Oklahoma. Dept. 859

TILIA The genus Tilia in Oklahoma. Proc. Okla. Acad. Sci. 50: 104-105. {SE 394

TOLERANCE Salinity tolerance of marsh plants in Kansas and Oklahoma. Ecolog 791

TOPOGRAPHIC Edaphic and topographic effects on forest communities in Payne 536

TOXIC Toxic plants of Oklahoma: Species causing phytodermatoses. J. Okla. V 94
TOXIC Toxic plants of Oklahoma: Locoweeds. J. Okla. Vet. Med. Assoc. 33: 13 95
TOXIC Toxic plants of Oklahoma: Oak, black locust, Kentucky coffee-tree, an 96
TOXIC Toxic plants of Oklahoma: Thornapples and nightshades. J. Okla. Vet. 97
TOXIC Toxic plants of Oklahoma: Bitterweeds and sneezeweeds. J. Okla. Vet. 98
TOXIC Toxic plants of Oklahoma: Ornamental houseplants. J. Okla. Vet. Med. 99
TOXIC Toxic plants of Oklahoma: Hemlocks. J. Okla. Vet. Med. Assoc. 34: 30- 100
TOXIC Toxic plants of Oklahoma: Ornamental shrubs. J. Okla. Vet. Med. Assoc 101
TOXIC Toxic plants of Oklahoma: Coffeweeds and sennas. J. Okla. Vet. Med. A 102
TOXIC Toxic plants of Oklahoma: Photosensitizing plants and vetch. J. Okla. 103
TOXIC Toxic plants of Oklahoma: Groundsels and rattlebox. J. Okla. Vet. Med 104
TOXIC Toxic plants of Oklahoma: Ferns and horsetails. J. Okla. Vet. Med. As 106
TOXIC Toxic plants of Oklahoma: Cyanogenic plants. J. Okla. Vet. Med. Assoc 107
TOXIC Toxic plants of Oklahoma: Milkweeds. J. Okla. Vet. Med. Assoc. 36: 74 108
TOXIC Toxic plants of Oklahoma: Species causing reproductive disorders. J. 109
TOXIC Toxic plants of Oklahoma: Species causing acute respiratory distress 110
TOXIC Toxic plants of Oklahoma: Cocklebur. J. Okla. Vet. Med. Assoc. 37: 63 111
TOXIC Toxic plants of Oklahoma: Cotton and cottonseed. J. Okla. Vet. Med. A 202
TOXIC Toxic plants of Oklahoma: Clovers. J. Okla. Vet. Med. Assoc. 36: 30-3 203

TOXINS Plant succession following destruction of vegetation by toxins from 827

TRANSITION Identifying the prairie-forest transition using remote sensing. 196

TRANSVAAL Plant succession on abandoned fields in central Oklahoma and in t 643

TRAVERTINE The ecology of Honey Creek: Temporal patterns of the travertine 596

TREE Injury and death or recovery of trees in a prairie climate. Ecol. Mono 20
TREE Comparative water relations of four species of upland deciduous trees 23
TREE The arborescent flora of Oklahoma. M. S. Thesis, Univ. Wyo., Laramie. 134
TREE The drowning of trees in Lake Carl Blackwell. Proc. Okla. Acad. Sci. 2 228
TREE Silvics of forest trees of the United States. Agr. Handbook No. 271, U 241
TREE Why prairies are treeless. Science N. S. 22: 55-56. {fire burning gras 249

TREE Notes on trees, shrubs, and vines in the Cherokee Nation. Trans. Kans. 266
TREE A nature study outline on trees and shrubs for the first six grades. M 290
TREE A preliminary survey of important factors which affect tree developmen 292
TREE Effect of silting on tree development in the flood plain of the North 293
TREE Relation of climatic conditions, soil characteristics, and tree develo 294
TREE Drought years in central Oklahoma from 1710 to 1959 calculated from an 295
TREE An analysis of selected Oklahoma upland forest stands including both o 375
TREE Acer grandidentatum in Oklahoma. Rhodora 46: 445-450. {SW Canadian Cad 411
TREE Forest trees of Oklahoma, 12th. edition. Okla. For. Div., State Dept. 414
TREE Trees and shrubs of the Southeastern Oklahoma Protective Unit. Proc. O 416
TREE Net production relations of two tree species in southern Oklahoma. Pro 424
TREE Roadside trees and shrubs of Oklahoma. Univ. Okla. Press, Norman. {flo 453
TREE The ligneous flora of Rich Mountain, Arkansas and Oklahoma. J. Arnold 527
TREE Diversity in tree species in Oklahoma upland forests. Ecology 52: 876- 632
TREE Factor analysis of tree distribution patterns in Oklahoma. Ecology 58: 656
TREE Shrub distribution patterns in Oklahoma with comparisons to those exhi 657
TREE The trees and shrubs of Oklahoma. Okla. Geol. Surv. Circ. 4: 7-41. {fl 666
TREE Trees. Yearbook of agriculture. U. S. Govt. Printing Office, Washingto 795

TRIFOLIUM Allium ampeloprasum (Liliaceae) and Trifolium vesiculosum (Legumi 757

TRINITY Differential selection in juniper populations from the Baum Limesto 282

TRIODIA A new Triodia from Oklahoma. Rhodora 40: 243-244. {flora} 227

TRITICEAE The generic concept and circumscription in the Triticeae: an end 221

ULMUS Vegetation response following winged elm (Ulmus alata Michx.) and oak 172
ULMUS An analysis of an elm-ash floodplain community near Norman, Oklahoma. 541

ULTISOL The Gulf Coastal Plain Ultisol and Alfisol ecosystem: surface geolo 502

UMBELLIFERAE A taxonomic study of the Umbelliferae in Oklahoma. M. S. Thesi 454

UNDERGROUND The detection and nature of functional underground connections 422

UNDERSTORY Vegetation response following winged elm (Ulmus alata Michx.) an 172
UNDERSTORY An analysis of selected Oklahoma upland forest stands including 375
UNDERSTORY Competitive and allelopathic suppression of understory by Oklaho 476

UNDISTURBED Causes of decreases in productivity in undisturbed tall grass p 613

UNMOWED A phenological comparison of mowed and unmowed tallgrass prairie si 488

UNRECLAIMED Revegetation of unreclaimed coal strip mines in Oklahoma. II. P 256
UNRECLAIMED Revegetation of unreclaimed coal strip-mines in Oklahoma. I. Ve 348

UPLAND Some factors influencing the frequency of bark lichens in north cent 6
UPLAND The effect of host specificity on the interspecific associations of 7
UPLAND Comparative water relations of four species of upland deciduous tree 23
UPLAND The woody vegetation of Bear's Glen, a Washington Irving stopover. P 40
UPLAND The Cross Timbers. Star Printery, Muskogee, OK. {upland forest tree} 238
UPLAND Geographic distribution of two species of oaks in Oklahoma in relati 284
UPLAND Some vegetation-environment relationships in the upland forests of O 344
UPLAND Biomass, annual net primary production and dynamics of six mineral e 345
UPLAND Woody vegetation of southeastern LeFlore County, Oklahoma, in relati 347
UPLAND Some vegetation-environment relationships in the upland forests of O 349
UPLAND Correlation analysis of rainfall and annual ring index of central Ok 350
UPLAND Biomass, annual net primary production, and dynamics of six mineral 351

UPLAND An analysis of selected Oklahoma upland forest stands including both 375
UPLAND The effects of periodic flooding on upland forest succession. M. S. 387
UPLAND The physiological ecology of upland and lowland Panicum virgatum. Ph 505
UPLAND Edaphic and topographic effects on forest communities in Payne Count 536
UPLAND The composition of a post oak forest in south-central Oklahoma. Sout 549
UPLAND A comparative study of upland and bottomland types of Panicum virgat 582
UPLAND An analysis of variation between upland and lowland switchgrass, Pan 583
UPLAND The microclimate of sugar maple stands in Oklahoma. Ecology 43: 19-2 601
UPLAND The upland forests of Oklahoma. Ecology 40: 593-608. {all forest} 621
UPLAND Diversity in tree species in Oklahoma upland forests. Ecology 52: 87 632
UPLAND Phytosociological analysis of Oklahoma upland forest species. Ecolog 633
UPLAND Factor analysis of tree distribution patterns in Oklahoma. Ecology 5 656
UPLAND Dispersion and co-dispersion in Oklahoma upland forests. Bull. Torre 678
UPLAND The comparative ecology of two canyons and an upland area in west ce 837

UREASE Soil enzymes in relation to old field succession: amylase, cellulase 531

VARIABLE-RADIUS An evaluation of the variable-radius and paired-tree method 619

VASCULAR A floristic study of the vascular plants of the Gypsum Hills and R 35
VASCULAR A preliminary survey of the vascular plants of Pottawatomie County 43
VASCULAR A preliminary list of the vascular plants of Pottawatomie County, 44
VASCULAR Vascular plants of Greer County, Oklahoma. M. S. Thesis, Univ. Okl 92
VASCULAR Vascular plants new for Oklahoma, Kansas, and Nebraska. Rhodora 73 429
VASCULAR A taxonomic study of the vascular plants indigenous to, or natural 455
VASCULAR Vascular plants of Pontotoc County, Oklahoma. Amer. Midl. Natur. 5 456
VASCULAR Vascular plants of southeastern Oklahoma from the Sans Bois to the 478
VASCULAR An outline for ecological life histories of herbaceous vascular hy 542
VASCULAR Primary production of vascular aquatic plants. Limnol. and Oceanog 545
VASCULAR Annotated checklist of the vascular plants of Little Sahara State 679
VASCULAR Additions to the vascular flora of Oklahoma. Rhodora 67: 191-193. 737
VASCULAR The vascular flora of Oklahoma – additions and comments. Rhodora 738
VASCULAR Additions to the vascular flora of Oklahoma – IV. Sida 12: 233-23 739
VASCULAR Additions to the vascular flora of Oklahoma. Sida 7: 361-368. {} 740
VASCULAR The genus Achillea. In: A synonimized checklist of the vascular fl 777
VASCULAR Vascular plants of the Oklahoma Ozarks. Ph. D. Diss., Okla. State 804
VASCULAR Some geographic relationships of the vascular flora of the Oklahom 825
VASCULAR A list of the vascular flora of the Oklahoma Ozarks. Proc. Okla. A 826

VEGETATION Allelopathic effects of Platanus occidentalis L. in relation to 17
VEGETATION The effect of cement-kiln dust on vegetation. M. S. Thesis, Univ 21
VEGETATION Vegetation-soil relationships in Flint Hills bluestem pastures. 27
VEGETATION Vegetation of the Salt Plains Wildlife Refuge, Jet, Oklahoma. Ph 33
VEGETATION The woody vegetation of Bear's Glen, a Washington Irving stopove 40
VEGETATION Standing state and cycling of nitrogen in soil-vegetation compon 61
VEGETATION Revegetation of abandoned fields in Kansas and Oklahoma. Amer. J 67
VEGETATION The vegetation of Oklahoma. Ecol. Monogr. 1: 99-188. {forest tre 79
VEGETATION Relationships of the woody vegetation of the Wichita Mountains W 84
VEGETATION Relationships of the woody vegetation of the Wichita Mountains W 85
VEGETATION Vegetation analysis of bobwhite habitat of northeastern Oklahoma 138
VEGETATION The effect of early spring burning on vegetation in buffalo wall 142
VEGETATION The relation of stock pond vegetation to geological formations i 145
VEGETATION Vegetation structure in a tallgrass prairie. M. S. Thesis, Univ. 146
VEGETATION Final report on vegetation and microenvironments of Platt Nation 171
VEGETATION Vegetation response following winged elm (Ulmus alata Michx.) an 172

VEGETATION Vegetational responses following winged elm and oak control in O 173
VEGETATION Investigations in erosion control and reclamation of eroded land 174
VEGETATION Evaluation of Payne County, Oklahoma farmlands and vegetation pa 177
VEGETATION The vegetation of the Wichita Mountains Wildlife Refuge. M. S. T 186
VEGETATION Seed hay for regrassing eroded land. Proc. Okla. Acad. Sci. 25: 206
VEGETATION Revegetation and gully-control experiments in the Red Plains Re 209
VEGETATION Interseeding and paraquat effects on central and eastern rangela 224
VEGETATION The vegetation of an eroding canyon in Canadian County, Oklahoma 245
VEGETATION The vegetation of an actively eroding canyon in Canadian County, 246
VEGETATION A vegetational analysis of the Phillips Agricultural Demonstrati 248
VEGETATION The effects of nitrogen fertilization on native vegetation under 250
VEGETATION The effects of one year's nitrogen fertilization on native veget 251
VEGETATION The natural revegetation of lead/zinc mine spoil in northeastern 255
VEGETATION Revegetation of unreclaimed coal strip mines in Oklahoma. II. Pl 256
VEGETATION Petroleum and surface vegetation. Proc. Okla. Acad. Sci. 10: 110 267
VEGETATION Effect of silt on natural vegetation and drainage in the flood p 297
VEGETATION A comparison of forest vegetation on orphan strip mines and adja 299
VEGETATION A floristic study of the woody vegetation of the North American 302
VEGETATION Vegetation resources including endangered and threatened species 312
VEGETATION Retrogression of native range vegetation in the sandy high plain 317
VEGETATION Vegetation in relation to slope exposure and geology in the Arbu 335
VEGETATION Some vegetation-environment relationships in the upland forests 344
VEGETATION Vegetational changes in a black willow forest over a 23-year per 346
VEGETATION Woody vegetation of southeastern LeFlore County, Oklahoma, in re 347
VEGETATION Revegetation of unreclaimed coal strip-mines in Oklahoma. I. Veg 348
VEGETATION Some vegetation-environment relationships in the upland forests 349
VEGETATION Changes in vegetation on stock pond dams near Norman, Oklahoma. 366
VEGETATION Vegetation and soil conditions of prairie and pasture plots in c 367
VEGETATION The vegetation of stock pond dams in central Oklahoma. Amer. Mid 371
VEGETATION Literature on the vegetation of Oklahoma. Proc. Okla. Acad. Sci. 372
VEGETATION Potential natural vegetation of the conterminous United States. 388
VEGETATION The vegetation of Muskogee County, Oklahoma. Amer. Midl. Natur. 409
VEGETATION The vegetation of the Caddo County Canyons, Oklahoma. Ecology 20 410
VEGETATION Allelopathic effects of Celtis laevigata Willd. as related to pa 417
VEGETATION Ecological factors affecting the distribution of woody vegetatio 420
VEGETATION Revegetation of coal stripped land near Henryetta, Oklahoma. Pro 439
VEGETATION A floristic study of the native or naturalized vegetation of Was 459
VEGETATION Literature on the vegetation of Oklahoma, 1964-1975. Proc. Okla. 481
VEGETATION Additions to the literature on the vegetation of Oklahoma. Proc. 484
VEGETATION A vegetation analysis of a pimpled prairie in northeastern Oklah 491
VEGETATION Revegetation of eroded lands in the High Plains of Texas and Okl 508
VEGETATION Some experiments and observations on revegetation of Oklahoma ra 515
VEGETATION Plant collections representative of some typical plant communiti 517
VEGETATION Plant collections representative of some typical communities in 518
VEGETATION The ecology of the western Oklahoma salt plains. Pub. Univ. Okla 519
VEGETATION Vegetation of an abandoned prairie dog town in tall-grass prairi 523
VEGETATION Soil factors preventing revegetation of a denuded area near an a 532
VEGETATION Vegetational changes in a black willow forest over a four year p 553
VEGETATION An analysis of the vegetation of buffalo wallows in relation to 579
VEGETATION Relationships of vegetation and environment in buffalo wallows. 581
VEGETATION The vegetation regions of the prairie province. Bot. Gaz. 25: 38 585
VEGETATION Allelopathic effects of Sporobolus pyramidatus on vegetational p 589
VEGETATION Allelopathic effects of Sporobolus pyramidatus on vegetational p 590

VEGETATION Correlation between vegetation and geologic formations in Oklaho 597
VEGETATION Vegetation of Beaver's Bend State Park. Okla. Geol. Surv. Guide 602
VEGETATION The vegetation of the Mesa de Maya region of Colorado, New Mexic 639
VEGETATION Natural revegetation of abandoned farmland in the central and so 650
VEGETATION Natural vegetation as an indicator of the capabilities of land f 667
VEGETATION The natural vegetation of the Great Plains region. Ann. Assoc. A 668
VEGETATION Natural vegetation. Atlas of American Agriculture, Sec. E: 1-29. 669
VEGETATION Vegetation of the Woods County, Oklahoma sand dunes. Ph. D. Thes 677
VEGETATION The pattern of retrogression of native vegetation in north centr 686
VEGETATION The structure and function of ten western North American grassla 687
VEGETATION Pattern of retrogression of native vegetation in north central O 688
VEGETATION Effect of disturbance on microclimate and vegetation in tall gra 693
VEGETATION Revegetation of strip mine spoils. M. S. Thesis, Okla. State Uni 702
VEGETATION The relation of forest vegetation to soils and geology in the gu 729
VEGETATION A study of revegetation in the floodplain of the South Canadian 807
VEGETATION The vegetation of the lower levels of the South Canadian River i 808
VEGETATION Plant succession following destruction of vegetation by toxins f 827
VEGETATION A survey of the soil algae of two vegetation types in central Ok 847

VEGETATIVE The influence of varying light intensities on the vegetative gro 49
VEGETATIVE Vegetative habit and reproduction of Desmidium grevillii (Kütz) 154
VEGETATIVE The identification of some of the more common native Oklahoma gr 301
VEGETATIVE Forage production, vegetative composition, and plant vigor in re 304
VEGETATIVE The significance of vegetative reproduction in Quercus. Madroño 490
VEGETATIVE A comparison of the predominant soil microflora in eight vegetat 706

VERATRUM Veratrum woodii (Melanthium woodii). Status report. Endangered Spe 742

VERBENA Taxonomic studies in the Verbena stricta complex. Ph. D. Thesis, Un 37

VETCH Toxic plants of Oklahoma: Photosensitizing plants and vetch. J. Okla. 103

VINES Notes on trees, shrubs, and vines in the Cherokee Nation. Trans. Kans 266
VINES The increasing importance of vines in southern Oklahoma. Proc. Okla. 539
VINES A plant community dominated by vines. Proc. Okla. Acad. Sci. 45: 41-4 555

VIRGIN A comparison of the soil fungi of a virgin prairie and an abandoned 461
VIRGIN A comparison of the insects of a virgin prairie, a grazed pasture, a 537

WALLOWS A preliminary study of buffalo wallows in the vicinity of Norman, O 45
WALLOWS The effect of early spring burning on vegetation in buffalo wallows 142
WALLOWS An analysis of the vegetation of buffalo wallows in relation to env 579
WALLOWS The relationship of plant species heterogeneity to soil variation i 580
WALLOWS Relationships of vegetation and environment in buffalo wallows. Ame 581

WATER Comparative water relations of four species of upland deciduous trees 23
WATER Water quality and phytoplankton communities in Lake Thunderbird, Clev 47
WATER Hydrologic and nutrient relationships in a pine-hardwood forest in so 137
WATER Water fluctuations as a factor in the life of six of the higher plant 178
WATER Water fluctuation as a factor in the life of the higher plants of a 3 181
WATER Water relations of some grasses with phreatophytic properties. Ph. D. 205
WATER Water stress of tallgrass prairie plants in central Oklahoma. J. Rang 277
WATER A comparative study of the water relations of Quercus marilandica and 283
WATER Geographic distribution of two species of oaks in Oklahoma in relatio 284
WATER The effect of water potential upon the geographic distribution of Car 318
WATER Epiphytic diatoms as indicators of water quality in farm and experime 765
WATER Epiphytic diatoms as indicators of water quality in farm and experime 768

WATERMILFOIL Eurasian watermilfoil – A new menace to Oklahoma waters. Proc 341

WATERSHEDS Biotic type mapping of Oklahoma watersheds. Proc. Okla. Acad. Sc 521
WATERSHEDS Sediment-precipitation-runoff relationships for the three epheme 638
WATERSHEDS Descriptive and hydrologic characteristics of three forested wat 801

WEATHER Weather patterns and ragweed production in Tulsa, Oklahoma. Ann. Al 87
WEATHER Ragweed pollen and weather patterns in Tulsa, Oklahoma. Bull. Ecol. 88
WEATHER An assessment of weather patterns and ragweed pollen in Tulsa over 89

WEED Weather patterns and ragweed production in Tulsa, Oklahoma. Ann. Aller 87
WEED Ragweed pollen and weather patterns in Tulsa, Oklahoma. Bull. Ecol. So 88
WEED An assessment of weather patterns and ragweed pollen in Tulsa over a f 89
WEED Toxic plants of Oklahoma: Locoweeds. J. Okla. Vet. Med. Assoc. 33: 13- 95
WEED Toxic plants of Oklahoma: Bitterweeds and sneezeweeds. J. Okla. Vet. M 98
WEED Toxic plants of Oklahoma: Coffeweeds and sennas. J. Okla. Vet. Med. As 102
WEED Toxic plants of Oklahoma: Milkweeds. J. Okla. Vet. Med. Assoc. 36: 74- 108
WEED Field surveys for noxious weeds in Oklahoma, Project 12-16-5-2301. Rep 477
WEED Germination of downy brome from southern Kansas, central Oklahoma, and 483
WEED Germination of weed seeds with reference to temperatures and storage c 493
WEED Phenology of spring weeds in Norman, Oklahoma. Southw. Natur. 1: 121-1 546
WEED Allelopathic inhibition of weed invasion by Andropogon scoparius Michx 713

WICHITA Relationships of the woody vegetation of the Wichita Mountains Wild 84
WICHITA Relationships of the woody vegetation of the Wichita Mountains Wild 85
WICHITA Some additions to Eskew's list of flowering plants of the Wichita M 90
WICHITA Grassland communities of the Wichita Mountains Wildlife Refuge. Ph. 162
WICHITA Influence of soils and parent materials on grasslands of the Wichit 163
WICHITA The vegetation of the Wichita Mountains Wildlife Refuge. M. S. Thes 186
WICHITA The flowering plants of the Wichita National Forest. M. S. Thesis, 216
WICHITA The flowering plants of the Wichita Mountains Wildlife Refuge. Amer 217
WICHITA The occurrence of persimmon and buckeye in the Wichita Mountains Wi 285
WICHITA A bibliography of the Wichita Mountains of Oklahoma. Great Plains J 286
WICHITA Geology and petrology of the Wichita Mountains. Okla. Geol. Surv. B 319
WICHITA Grazing use checks on the Wichitas. J. Range Manage. 16: 142-145. { 378
WICHITA A preliminary report on the flora of the Wichita Mountains. M. S. T 492
WICHITA Influence of a wildfire in the Wichita Mountains Wildlife Refuge. E 554

WILDFIRE Influence of a wildfire in the Wichita Mountains Wildlife Refuge. 554

WILDLIFE Vegetation of the Salt Plains Wildlife Refuge, Jet, Oklahoma. Ph. 33
WILDLIFE Bottomland hardwoods of eastern Oklahoma: A special study of their 72
WILDLIFE Comparative studies of the animal communities of oak-hickory fores 76
WILDLIFE Relationships of the woody vegetation of the Wichita Mountains Wil 84
WILDLIFE Relationships of the woody vegetation of the Wichita Mountains Wil 85
WILDLIFE Some additions to Eskew's list of flowering plants of the Wichita 90
WILDLIFE The effect of salinity on the distribution of algae at the Salt Pl 113
WILDLIFE Growth of ten species of preferred deer browse plants under variou 132
WILDLIFE The life-form concept and mule deer habitat in Oklahoma. Proc. Okl 135
WILDLIFE Vegetation analysis of bobwhite habitat of northeastern Oklahoma r 138
WILDLIFE Grassland communities of the Wichita Mountains Wildlife Refuge. Ph 162
WILDLIFE Influence of soils and parent materials on grasslands of the Wichi 163
WILDLIFE Components of the habitat of the bobwhite quail in Payne County, O 175
WILDLIFE The vegetation of the Wichita Mountains Wildlife Refuge. M. S. The 186
WILDLIFE An evaluation of ground nesting by mourning doves in northwestern 193

WILDLIFE The flowering plants of the Wichita Mountains Wildlife Refuge. Ame 217
WILDLIFE The occurrence of persimmon and buckeye in the Wichita Mountains W 285
WILDLIFE Habitat preferences among representative wintering and breeding bi 353
WILDLIFE The life-form concept applied to prairie chicken habitat in Oklaho 354
WILDLIFE 1963. Identification and analysis of lesser and greater prairie ch 356
WILDLIFE Habitat used by lesser prairie chickens for feeding related to sea 357
WILDLIFE Prairie dogs, whitefaces, and blue grama. Wildl. Monogr. 3: 1-78. 386
WILDLIFE The effect of land use on nest sites selected by Mississippi kites 421
WILDLIFE Influence of a wildfire in the Wichita Mountains Wildlife Refuge. 554
WILDLIFE Flora and fauna of Oklahoma abandoned mine lands. Okla. Biol. Surv 654
WILDLIFE Model fish and wildlife plan for a surface coal mine in Oklahoma. 655
WILDLIFE Wildlife and roadside erosion in central Oklahoma. Proc. Okla. Aca 694
WILDLIFE Habitat description and the life-form concept. Proc. Okla. Acad. S 707
WILDLIFE National Wildlife Refuges – What future? Ecosystems 6: 4-11. {} 727
WILDLIFE Habitat types used by the bobwhite quail and other wildlife in sou 772

WILLOW Vegetational changes in a black willow forest over a 23-year period. 346
WILLOW The composition of a black willow forest in south central Oklahoma. 547
WILLOW Vegetational changes in a black willow forest over a four year perio 553

WIND Wind patterns and the seasonal occurrence of Juniperus pollen in Tulsa 402

WINGED ELM Vegetation response following winged elm (Ulmus alata Michx.) an 172
WINGED ELM Vegetational responses following winged elm and oak control in O 173

WINTER Differential response of woody and herbaceous species to summer and 9
WINTER Distribution and identification of woody plants of Oklahoma in the w 86
WINTER Habitat preferences among representative wintering and breeding bird 353
WINTER Winter burning in central Oklahoma grassland. Ecology 38: 520-522. { 370
WINTER Germination of downy brome from southern Kansas, central Oklahoma, a 483
WINTER Preliminary observations on mourning doves wintering in southwestern 487
WINTER Some effects of winter burning on a moderately grazed pasture. Ecolo 560

WISTER A pre-impoundment biological survey of Wister Reservoir, Oklahoma. M 279

WOODED Grass stand establishment on rough, rocky, wooded lands in eastern O 423
WOODED Establishment of warm and cool season grass pastures on wooded sites 664

WOODLAND The short grass plains and post oak-blackjack woodland of Oklahoma 56
WOODLAND A quantitative comparison of the soil algae of central Oklahoma pr 213
WOODLAND Woodland of Indian Territory. U. S. Geol. Surv. Rep. No. 21: 603-6 235
WOODLAND A study of the environment and floristic composition of an oak-hic 426
WOODLAND Oklahoma forests and woodlands. Proc. Okla. Acad. Sci. 30: 3-10. { 684

WOODY Differential response of woody and herbaceous species to summer and w 9
WOODY Distribution and control of several woody plants in Texas and Oklahom 25
WOODY The woody vegetation of Bear's Glen, a Washington Irving stopover. Pr 40
WOODY Relationships of the woody vegetation of the Wichita Mountains Wildli 84
WOODY Relationships of the woody vegetation of the Wichita Mountains Wildli 85
WOODY Distribution and identification of woody plants of Oklahoma in the wi 86
WOODY Reproduction of woody plants in the absence of grazing. Proc. Okla. A 119
WOODY A floristic study of the woody vegetation of the North American Cross 302
WOODY Woody vegetation of southeastern LeFlore County, Oklahoma, in relatio 347
WOODY Ecological factors affecting the distribution of woody vegetation nea 420
WOODY The detection and nature of functional underground connections betwee 422
WOODY The phenology of woody plants in Norman, Oklahoma during the spring s 544
WOODY The woody plants of Oklahoma. Ph. D. Thesis, Univ. Calif., Berkeley. 683

WOODY Atlas of the woody plants of Oklahoma. Okla. Biol. Surv., Norman. {tr 845

ZINC Plant communities and the potential for ecotypic differentiation on le 254
ZINC The natural revegetation of lead/zinc mine spoil in northeastern Oklah 255
ZINC Soil factors preventing revegetation of a denuded area near an abandon 532
ZINC Plant succession following destruction of vegetation by toxins from a 827